The Medieval Floortiles of Herefordshire

About Access Archaeology

Access Archaeology offers a different publishing model for specialist academic material that might traditionally prove commercially unviable, perhaps due to its sheer extent or volume of colour content, or simply due to its relatively niche field of interest. This could apply, for example, to a PhD dissertation or a catalogue of archaeological data.

All *Access Archaeology* publications are available as a free-to-download pdf eBook and in print format. The free pdf download model supports dissemination in areas of the world where budgets are more severely limited, and also allows individual academics from all over the world the opportunity to access the material privately, rather than relying solely on their university or public library. Print copies, nevertheless, remain available to individuals and institutions who need or prefer them.

The material is refereed and/or peer reviewed. Copy-editing takes place prior to submission of the work for publication and is the responsibility of the author. Academics who are able to supply print-ready material are not charged any fee to publish (including making the material available as a free-to-download pdf). In some instances the material is type-set in-house and in these cases a small charge is passed on for layout work.

Our principal effort goes into promoting the material, both the free-to-download pdf and print edition, where *Access Archaeology* books get the same level of attention as all of our publications which are marketed through e-alerts, print catalogues, displays at academic conferences, and are supported by professional distribution worldwide.

The free pdf download allows for greater dissemination of academic work than traditional print models could ever hope to support. It is common for a free-to-download pdf to be downloaded hundreds or sometimes thousands of times when it first appears on our website. Print sales of such specialist material would take years to match this figure, if indeed they ever would.

This model may well evolve over time, but its ambition will always remain to publish archaeological material that would prove commercially unviable in traditional publishing models, without passing the expense on to the academic (author or reader).

Archaeopress Access Archaeology

The Medieval Floortiles of Herefordshire

Julie Bowen

Access Archaeology

Archaeopress Access Archaeology

ARCHAEOPRESS PUBLISHING LTD
Summertown Pavilion
18-24 Middle Way
Summertown
Oxford OX2 7LG
www.archaeopress.com

ISBN 978-1-78969-513-7
ISBN 978-1-78969-514-4 (e-Pdf)

Contents

Acknowledgements

I am indebted to the National Museum of Wales for allowing use of the drawings and tile descriptions published in J. M. Lewis' exhaustive book on The Medieval Tiles of Wales.

I am grateful to Laurence Keen, the Director of the Census of Medieval Tiles in Britain, for his support in this endeavour, and for supplying unpublished reports on medieval floortiles found at Blackfriars, Gloucester, and at Worcester Cathedral. Mr Keen also gave me access to the unpublished drawings by Alan Hunt of the Much Marcle tiles.

With some of the heraldic shields I have drawn on the work done by Tony Wilmott, published in the Vince and Wilmott paper on 'A lost tile pavement at Tewkesbury Abbey' published in the Antiquaries Journal 1991. My thanks to Mr Wilmott for allowing me use of his drawings.

Hilary White kindly provided me with some notes she had made on medieval floortiles whilst she was part of the Sites and Monuments Record staff at Hereford. Her information on the Netherwood tiles at Thornbury was particularly helpful, and I am grateful to Ivan Turner who showed me his private collection of floor tiles from Netherwood.

It was Hilary White and Duncan Brown who had worked on the George Gilbert Scott tiles from Leominster Old Priory previously, and Duncan Brown allowed use of his earlier drawings.

Lynn Harper of Monmouth Archaeology permitted use of her drawings of some of the Abbey Dore tiles.

My thanks to Church and Site Archaeological Services for allowing me access to the floortiles retrieved from their watching briefs, in particular at Dilwyn and Breinton.

I am grateful to the Hereford Museum staff for granting me access to search for medieval floortiles within their Collections. Judy Stephenson identified the possible use of a barrel lock key in the production of a unique tile from Leominster Old Priory.

There were three locations, secular buildings now in private ownership, which house medieval floortiles, and I am most grateful to the owners of Wigmore Abbey; Chapel Farm, Wigmore; and Deans Place, Yatton, for allowing access to their tiles. Dr Patrick Brown helped with the identification of certain tiles from Deans Place and provided the main photograph.

Most of the tiles which appear to be unique to Herefordshire are illustrated by the author's own photographs and those of Andrew Stait. Other individual photographs are credited within the text.

Introduction

It had been my original intention to research the decorated medieval floortiles in Herefordshire in order to better understand the dating and distribution of the tiles and the links between the sites at the time the tiles were originally laid.

However, circumstances have dictated that this book is only the first step along that road, in that this is simply a survey, inevitably a snapshot in time, of the decorated medieval floortiles visible in the county of Herefordshire in 2013. I have also included a few instances where records exist of medieval floortiles having been found, but the actual tiles are no longer available. I hope that it will provide a foundation for others to undertake further research.

I am of course in the shadow of the late Alan Vince and his exhaustive studies of the ceramics industry in the Severn valley. His interest was primarily with the analysis of the fabric of the tiles, its origin, and the trade routes. However, the vast majority of the medieval tiles in Herefordshire are laid as flooring, albeit not in their original setting, and the fabric of the tile is not available for analysis. I can only work with the designs.

In this book I have not included plain glazed tiles, as these are difficult to date, and unhelpful in suggesting links between sites. It should be noted perhaps that a significant number of surviving tiles feature heraldic shields, but we should be wary of the assumption that these shields formed such a large proportion of the original tiles laid. It is likely that the workmen salvaging these tiles, often during Victorian restorations of the churches, considered the shields to be of more importance than the foliate designs.

It is impossible to look at Herefordshire in isolation, particularly with such large and well-documented collections of floortiles close by, notably at Great Malvern, Worcester, and Gloucester. Where other sites have tiles of the same design I have tried to list only those locations within a 20-mile radius of Herefordshire.

As the decorated tiles are now to be found in often dark buildings where photography is difficult, where there is a drawing available of the design then this is what I have used as this gives a clearer image. The photographs were taken mainly by the author and Andy Stait; those taken by Laurence Keen and Dr Patrick Brown are referenced in the text.

Given Herefordshire's shared border with the Welsh counties, there are many similar designs to those in J. M. Lewis' exhaustive book 'The Medieval Tiles of Wales'; these were drawn by Jeremy Leech and are referenced throughout the book. Laurence Keen produced the drawings of the floortiles from Blackfriars, Gloucester; Worcester Cathedral; and Netherwood. The heraldic shield tiles were drawn by Tony Wilmott: Lynn Harper produced some of the Abbey Dore drawings: Duncan Brown the Leominster Old Priory Gilbert-Scott tiles; the other Leominster Old Priory tile drawings and those of Dilwyn are the author's own. A few are derivatives from Elizabeth Eames' Catalogue of the British Museum tiles.

I have therefore listed the main sites alphabetically, and detailed under each site name the tiles visible at that location. Obviously some places (such as Stretton Sugwas and Croft) have a large number of tiles of the same design, but I have still listed all of them at both locations so that a reader selecting only one discrete site will still find a complete record within that chapter.

I am presenting a census of what is visible, or available in the record; it is as comprehensive as I can make it but I do not pretend it is complete. I hope that others will find it a useful basis for further research.

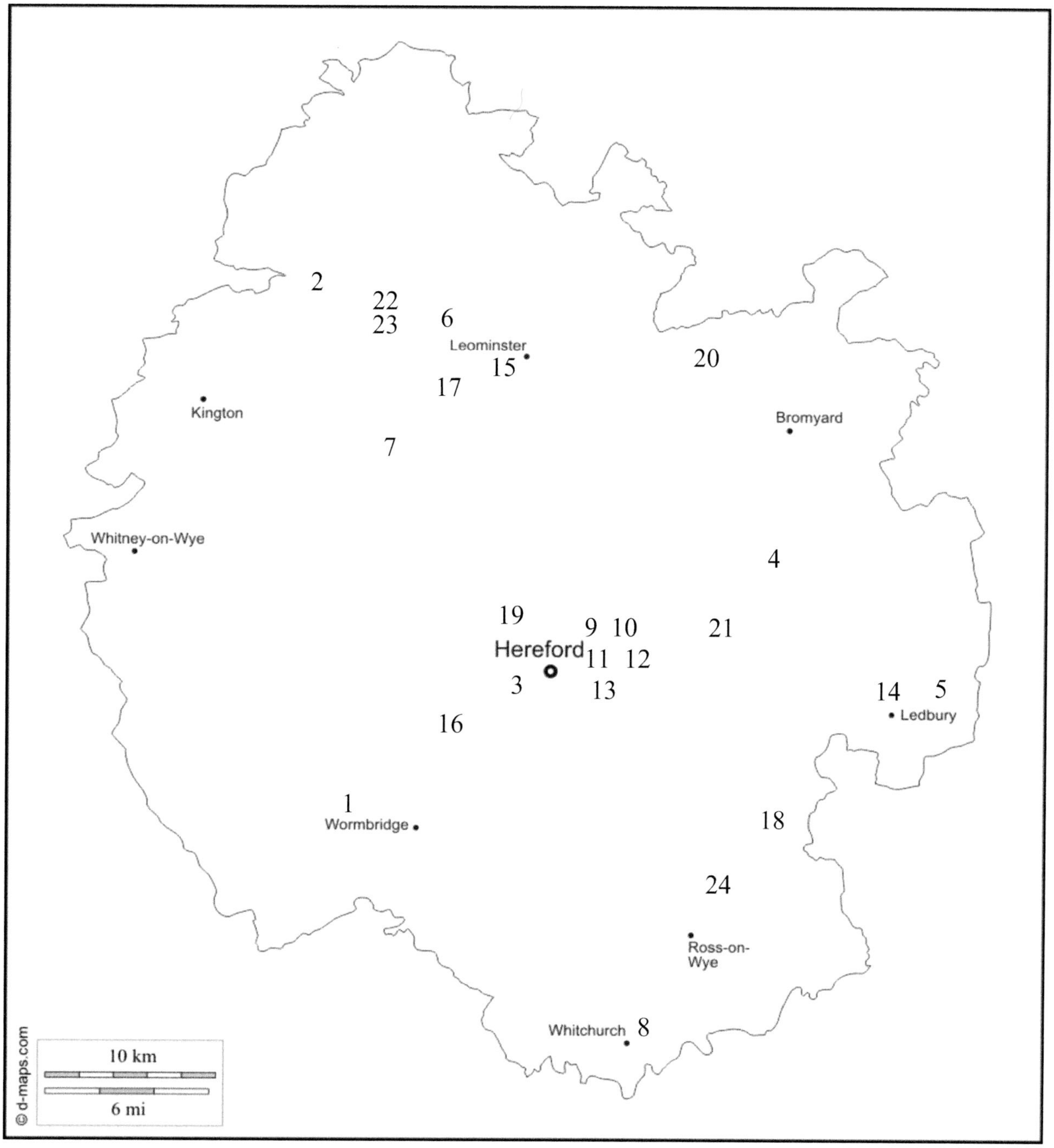

KEY:

1 Abbey Dore	7 Dilwyn	13 Hereford Museum	19 Stretton Sugwas
2 Brampton Bryan	8 Goodrich	14 Ledbury	20 Thornbury
3 Breinton	9 Hereford All Saints	15 Leominster	21 Westhide
4 Castle Frome	10 " Booth Chapel	16 Madley	22 Wigmore Abbey
5 Colwall	11 " Lady Chapel	17 Monkland	23 Wigmore Chapel Farm
6 Croft	12 " Eign Gate	18 Much Marcle	24 Yatton

Abbey Dore – Dore Abbey

There is inevitable confusion here, but I will refer to the abbey as Dore Abbey which lies within the community of Abbey Dore.

The medieval floortiles found here have been extensively reported in Alan Vince's chapter in the excellent book 'A Definite History of Dore Abbey' edited by Ron Shoesmith and Ruth E Richardson (Logaston 1997). I will give only a brief summary here, but would refer any reader to the above book for a very full description of the tiles.

Excavations by Roland Paul in 1901 recorded medieval floortiles, many of which were relaid to either side of the altar and others stored loose in a box in the vestry. The relaid tiles are predominantly lozenge-shaped relief tiles and two-colour heraldic shields, whereas the loose fragments are more random designs.

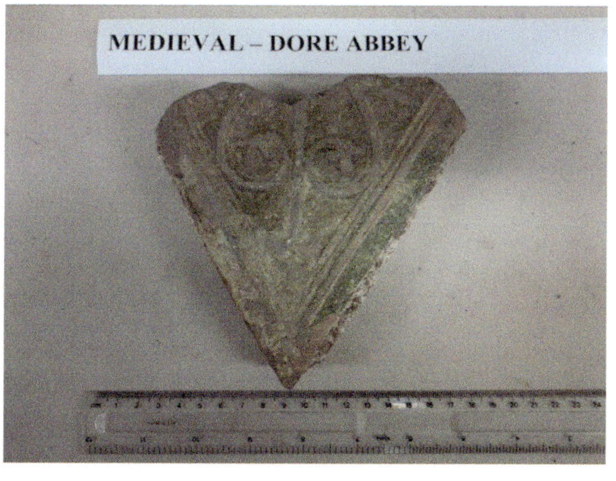

This green-glazed tile is in the collections of the Hereford Museum and would appear to be the same as that drawn in the Parker-Hore Collection (online Tileweb) (opposite).

Parker-Hore (Tileweb on line).

Tony Wilmott carried out extensive research into heraldic shields on floortiles, subsequently published in the Vince & Wilmott paper on 'A Lost Tile Pavement at Tewkesbury Abbey' (Antiquaries Journal 1991). He lists the following heraldic shields from Dore Abbey:

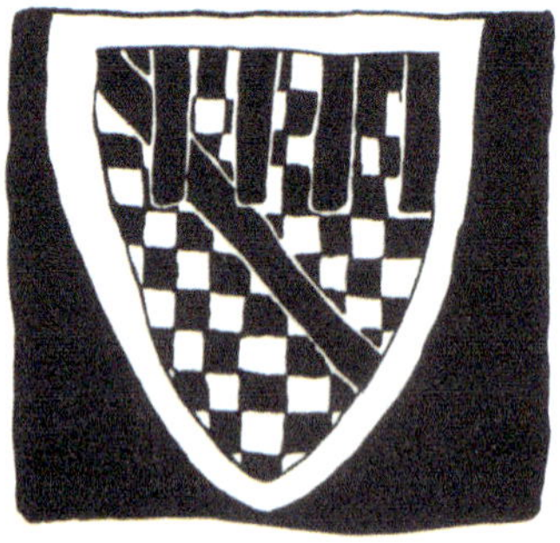

11

Chequy, a bend and a label of five points.
Clifford.

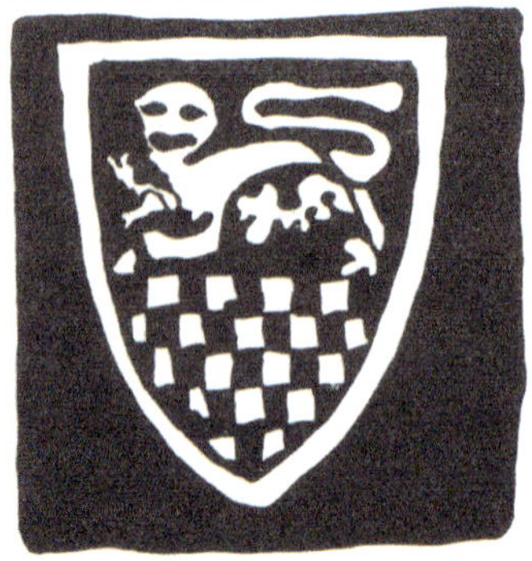

26

Chequy, on a chief a leopard.
Unknown.

34

Crusilly pate, a cross pate.
Berenger Moyne.

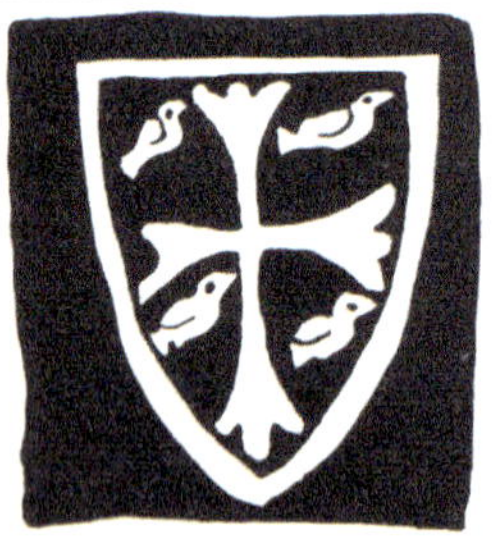

36

A cross patonce between five martlets.
Edward the Confessor (Here with 4 martlets – sometimes with a fifth martlet at the base of the cross.)
Also at Eign Gate, Hereford.

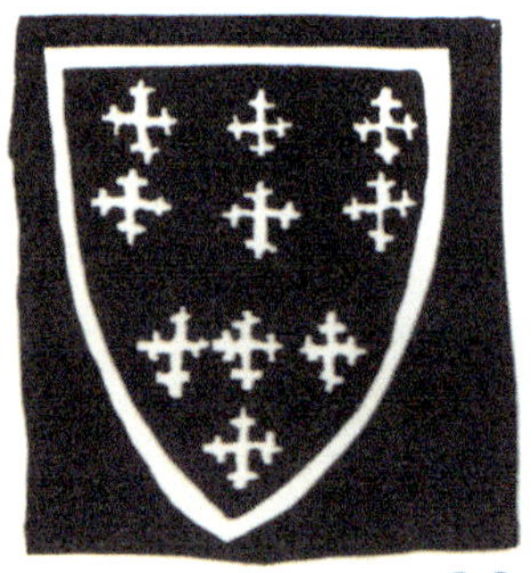

38

Crusilly of crosses crosslet.
Unknown.

39

Chequy, a fess.
Clifford.

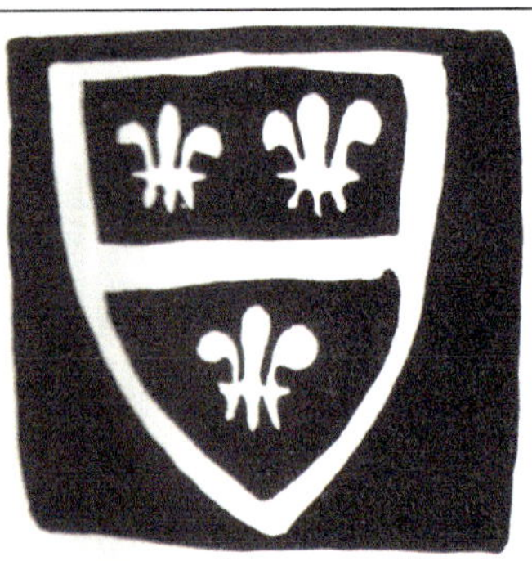

43

A fess between three fleur de lis
Edvin Ralph.
Also at Eign Gate, Hereford.

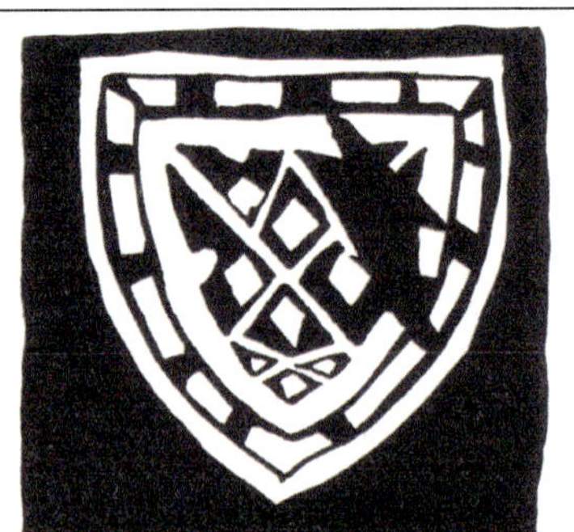

49

Masculy, a bordure gobonny.
Unknown.

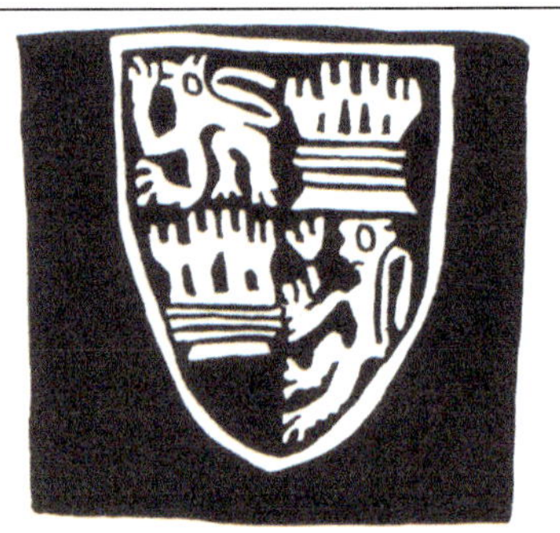

53

Quarterly, first and fourth, a lion rampant, second and third, a castle.
The arms of the Spanish house of Leon and Castile (after Edward I's marriage to Eleanor of Castile in 1254).
Also at Eign Gate, Hereford.

The following line drawings have been done by Lynn Harper, a member of Monmouth Archaeology.

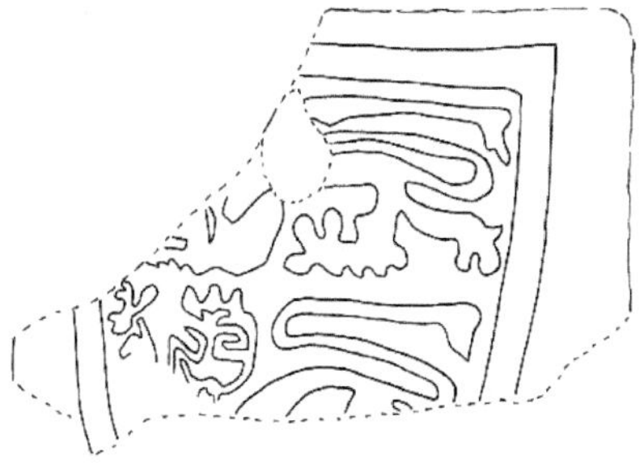

Three lions passant guardant
England.

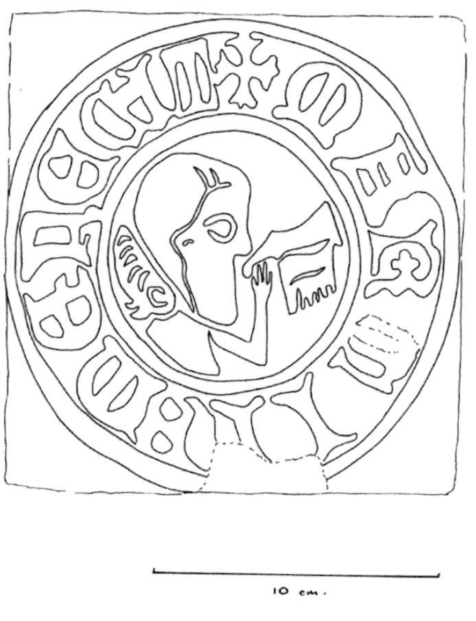

This tile has a circular band containing the Lombardic script "Martin me fecit" enclosing a central angel playing a trumpet.

A very worn example is in Hereford Cathedral, Booth Chapel.

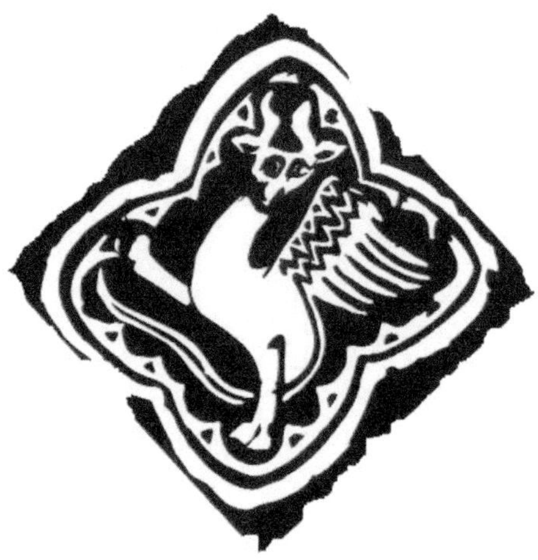

After Eames - British Museum 1389.

This winged and horned beast is also at Colwall, Eign Gate, Hereford, and at Blackfriars, Gloucester.

There are many fragments which probably form part of this 4- or 16-tile design of the Chertsey/Halesowen school, found at Leominster Old Priory. The Lombardic script, apparently in reverse, suggests the stamp makers were illiterate.

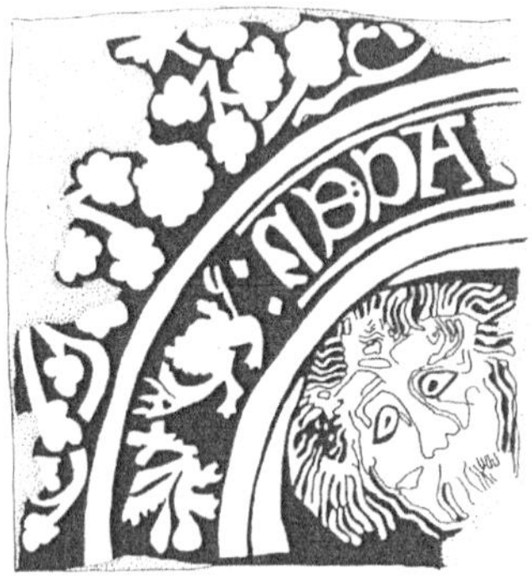

GS12
Leominster Old Priory.
Also Berrington St, Hereford.

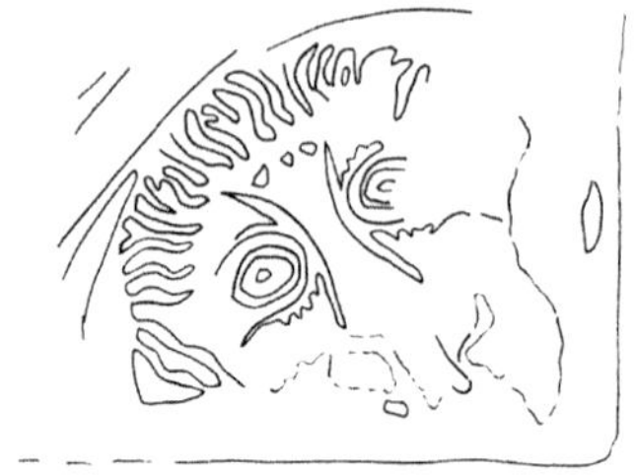

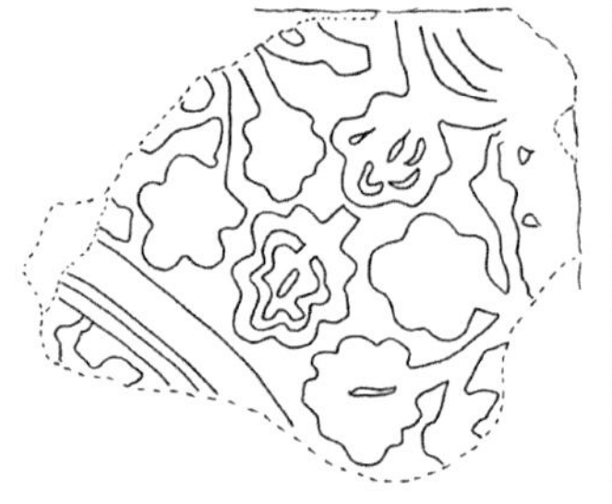

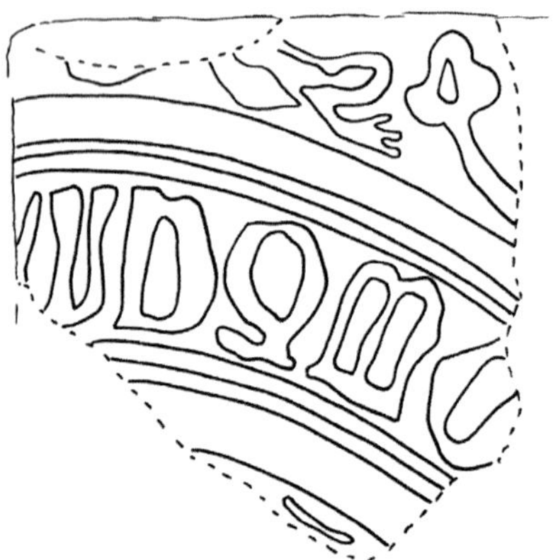

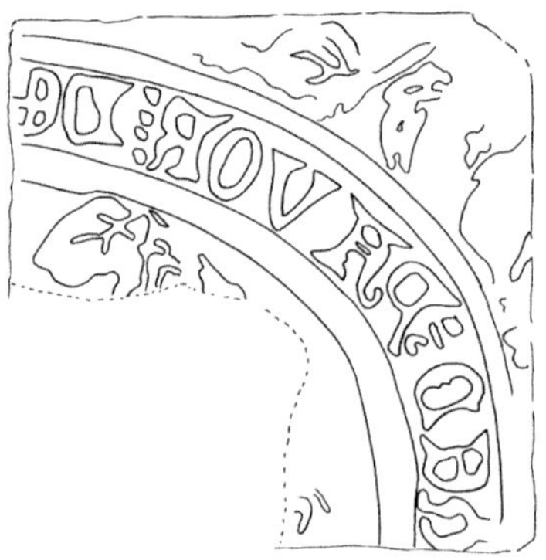

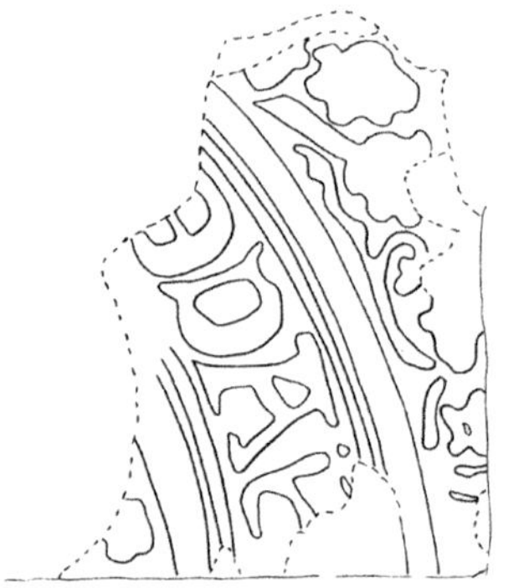

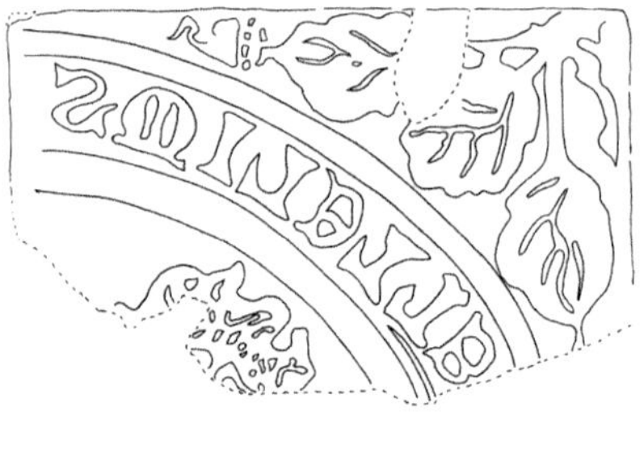

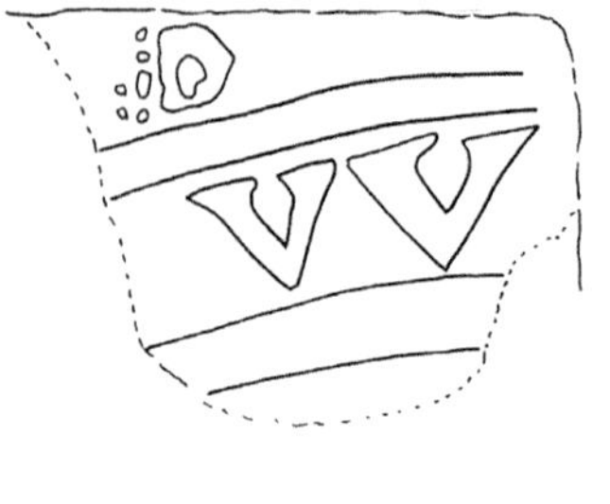

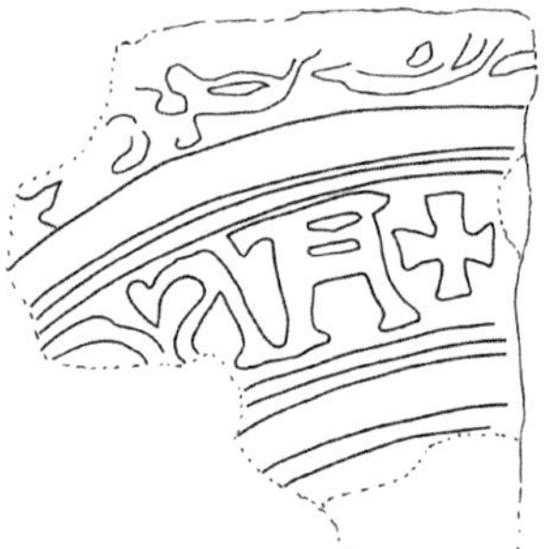

Brampton Bryan – Church of St Barnabas

The Guide Book to the Church and Castle tells us that Brampton was allotted by William the Conqueror to Ralph de Mortimer, who developed a 'little kingdom, known later as Walcheria, with its capital at Wigmore'. In 1179 we hear that 'Sir Hugh de Mortimer laid the first stone of the Abbey of Wigmore, Brian de Brampton laid the second and promised an hundred shillings.' There remained a close link between the two castles at Wigmore and Brampton Bryan until the Harleys (who owned both) deliberately destroyed Wigmore Castle to stop its use by Royalists in the Civil War.

The Norman Church of St Barnabas was destroyed during the Civil War and the present church therefore dates to circa 1656. Building material from the castle ruins was incorporated into the church, possibly including the medieval floortiles displayed today. In the south wall is a recess with a stone effigy of a lady, thought to date to the late fourteenth century, likely that of Margaret de Brampton who married the first Robert Harley. Behind this effigy are approximately 48 slip tiles, laid in three rows of sixteen, with a small fragment of dotted border tile at the front on the right.

These are an eclectic mix, with parallels found from Leominster Old Priory, Wigmore Abbey, and also St Michael's at Croft.

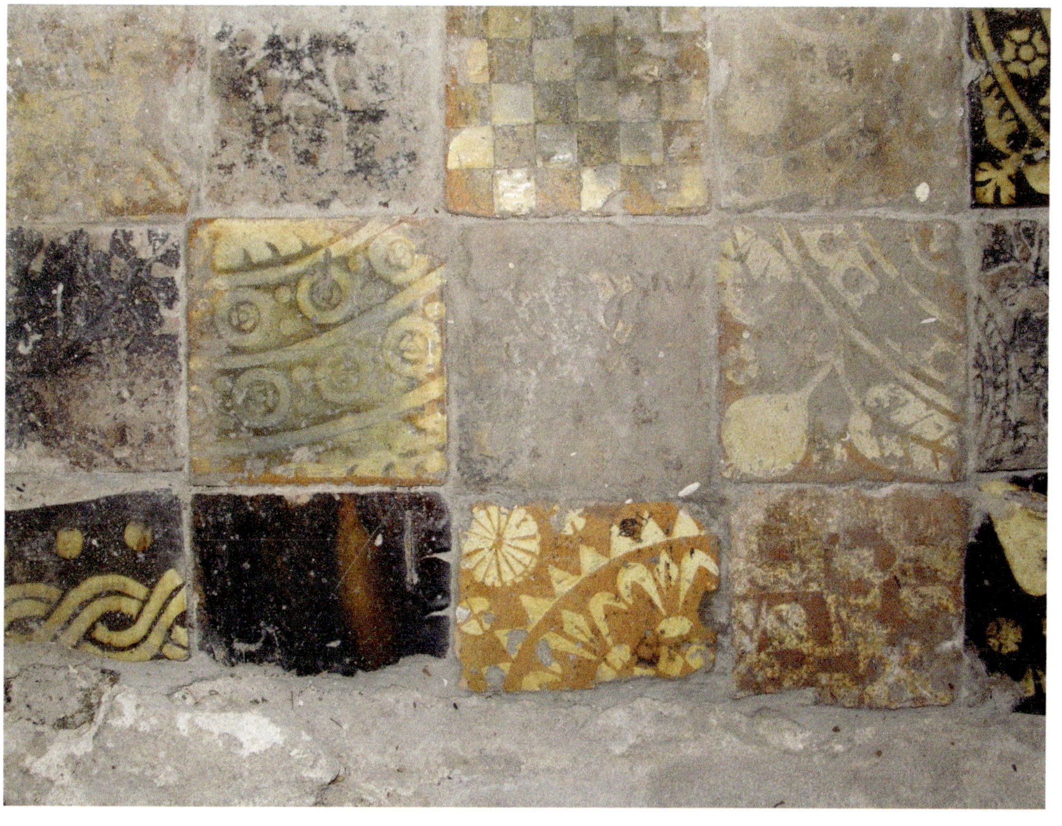

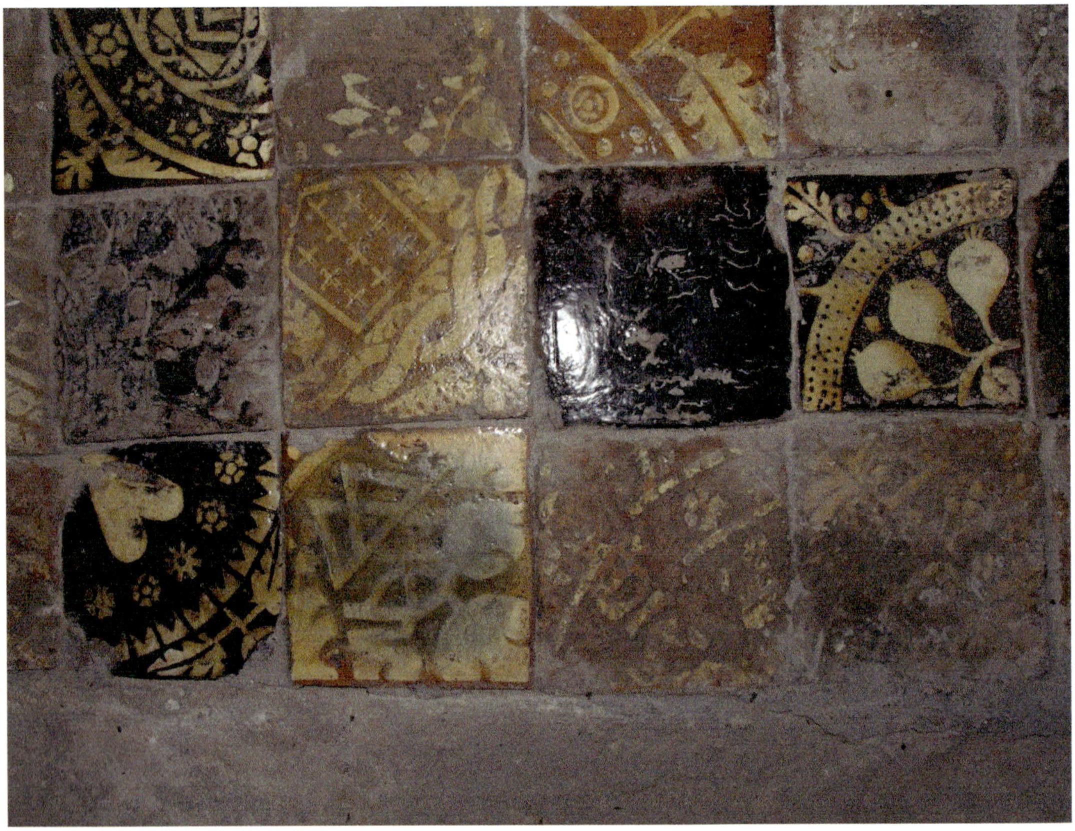

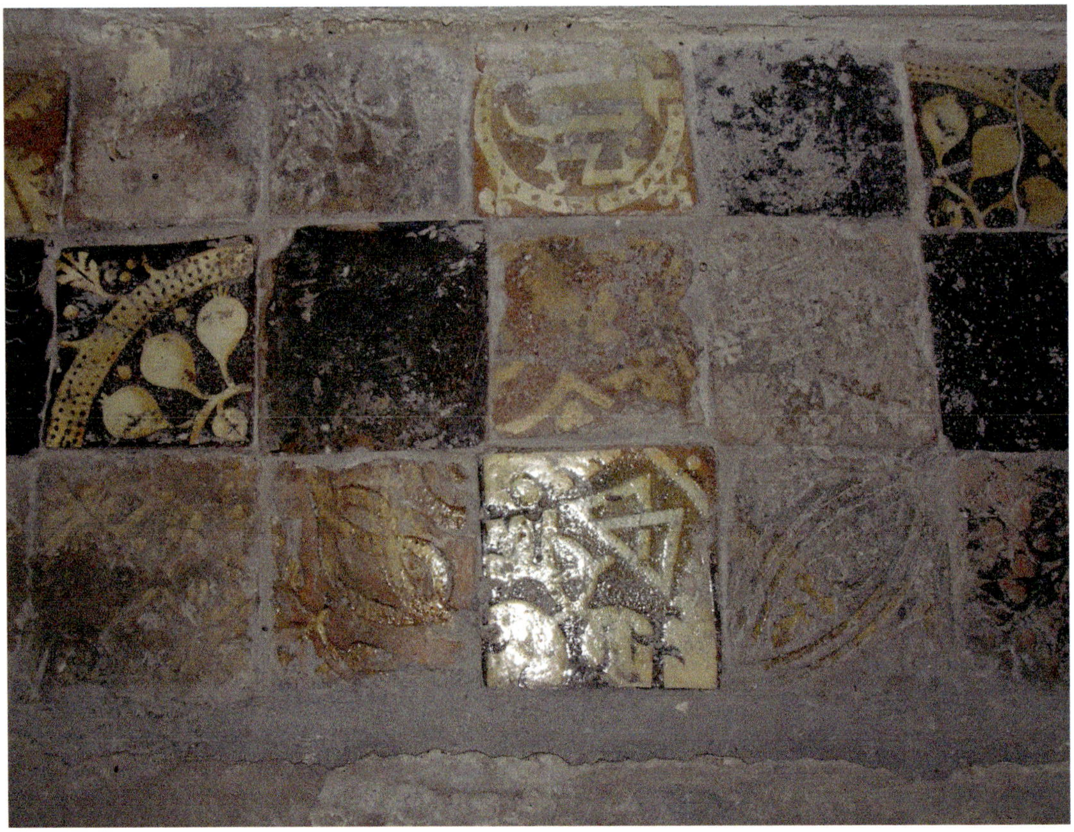

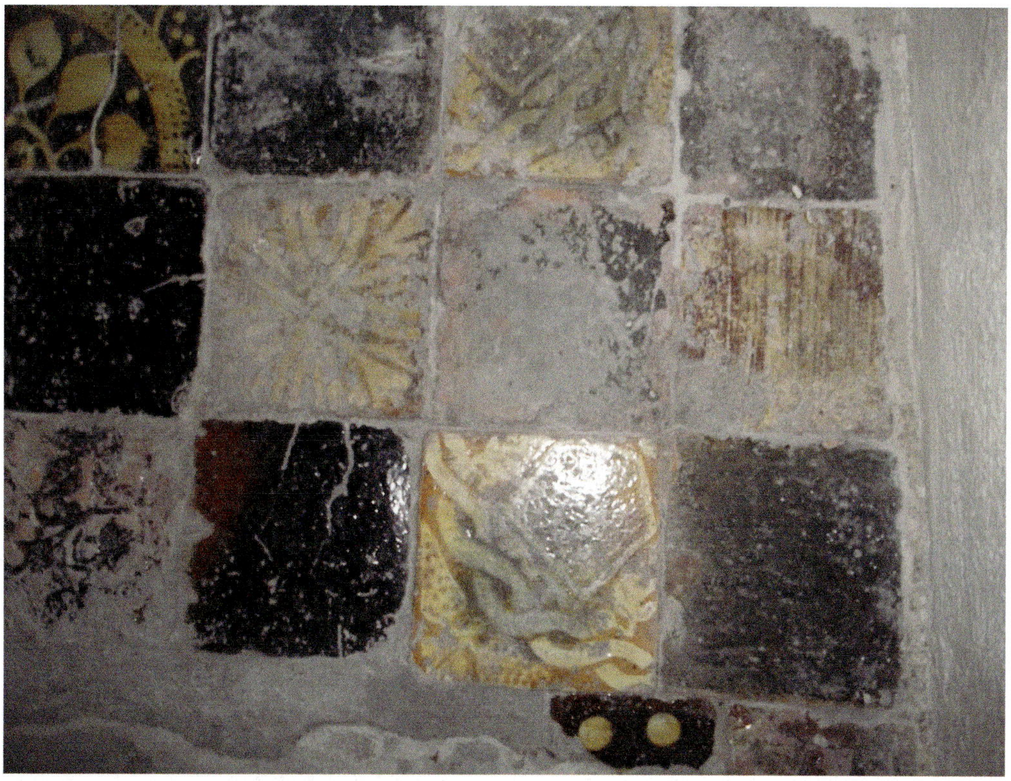

There are circa seven tiles with a plain very dark glaze, which Alfred Watkins commented on in 1920 during a visit by the Woolhope Club. 'Some of these tiles have a remarkable black glaze and Mr Alfred Watkins drew attention to its similarity to the glaze found on some of the pottery recently unearthed at Whitney (on Wye).'

There are perhaps another eight examples (very worn) of the impressed relief tiles found at Wigmore (JM Lewis 481), together with a single example of a design (480) from JM Lewis same Group 27.

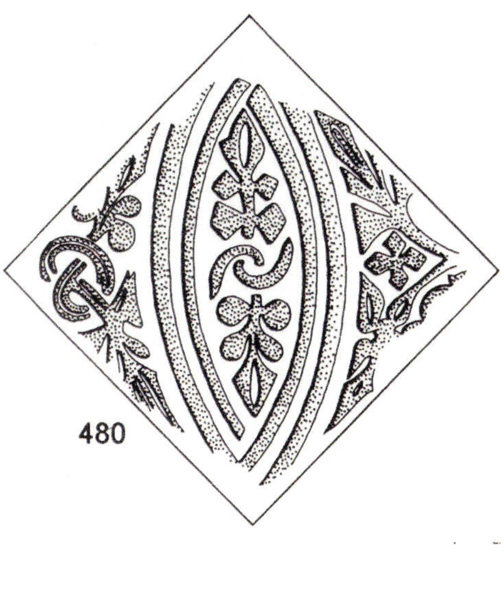

480

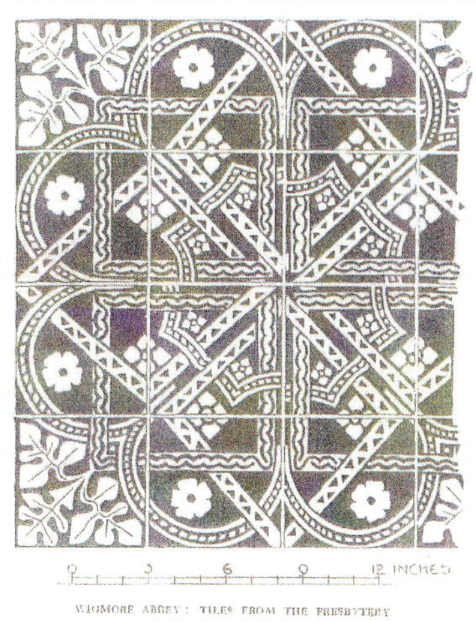

481 Corner tile (x8) of this design from Wigmore Abbey.

Many of the following tiles are also found at Leominster Old Priory (LOP) and at St Michael's, Croft.

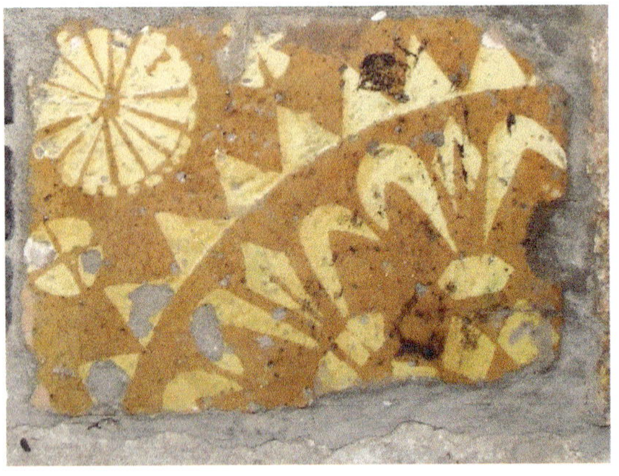

Also at LOP (BW.6) and St Michael's, Croft.

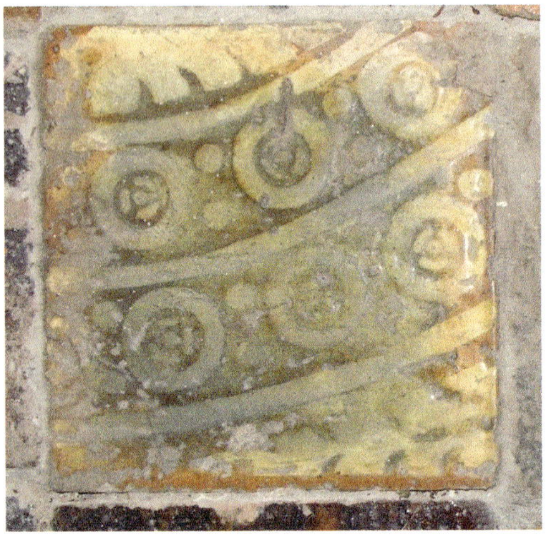

Also at LOP (BW.36) and St Michael's, Croft.

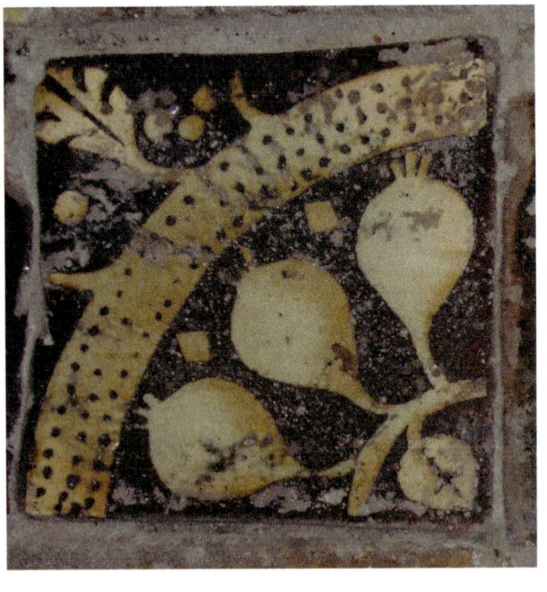

This Brampton Bryan example is a much darker glazed version of one found at Leominster Old Priory (BW.1). These have the identical small dots within the wide yellow band which, as Laurence Keen pointed out, would have been difficult to produce with anything but a metal (probably lead) stamp.

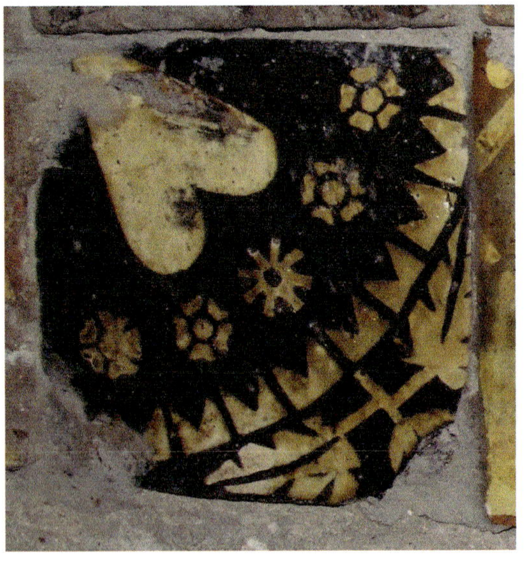

Also at Wigmore Abbey and St Michael's, Croft. This Brampton Bryan example is a particularly dark glaze, as is Croft.

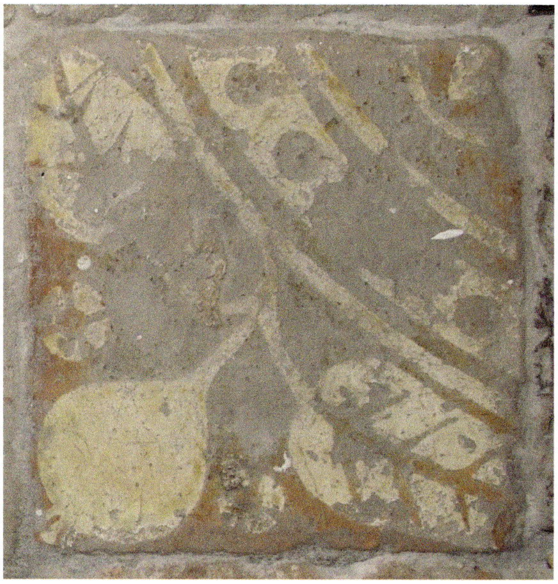

Also at St Michaels, Croft.

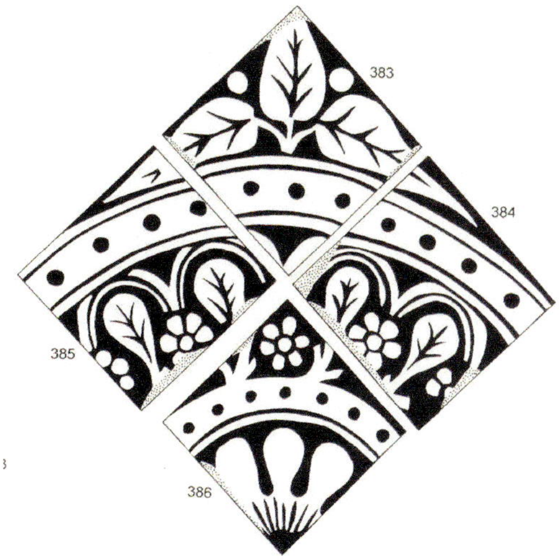

This Brampton Bryan tile (left) is similar to Canynges style JM Lewis 383 – but different corner fruit and different size dots.

(Ledbury, St Katherines and Deans Place, Yatton).

The following are drawings of tiles found at Leominster Old Priory which are also found at Brampton Bryan. I have used the drawings as they are clearer than the photographs.

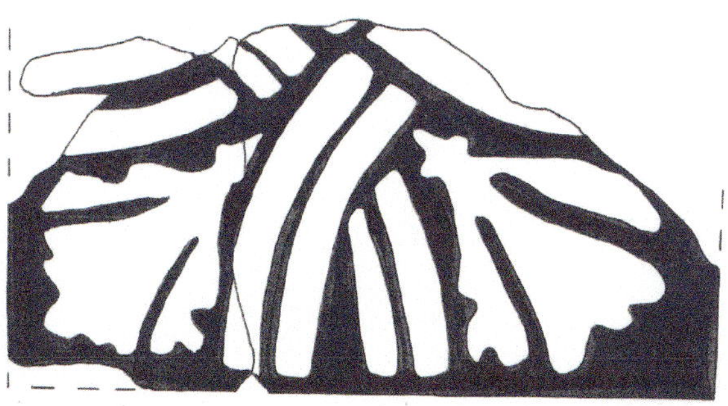

BW.9
Brampton Bryan, Leominster Old Priory, and St Michael's, Croft.

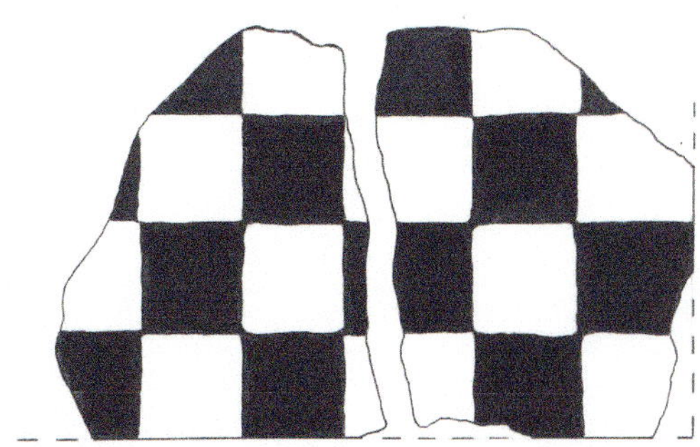

BW.27
Brampton Bryan, Leominster Old Priory, and St Michael's, Croft.

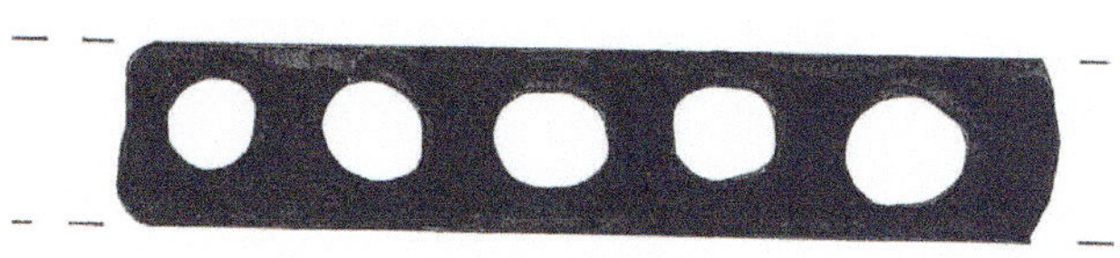

BW.53

A very unusual border strip.
Brampton Bryan, and Leominster Old Priory.

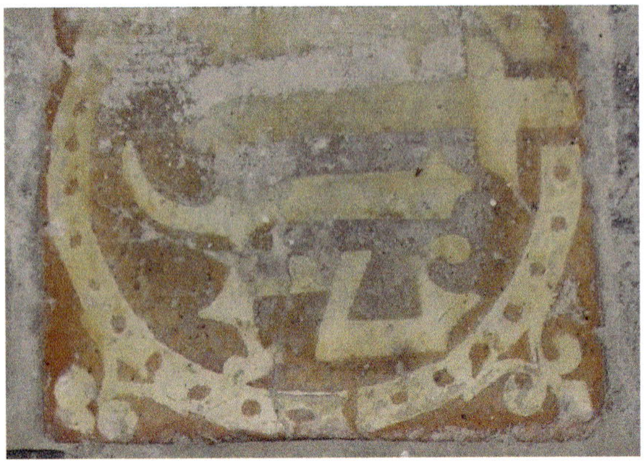

"ihc" within pelleted border; fleur-de lys at corners. Brampton Bryan, and St Michael's, Croft.

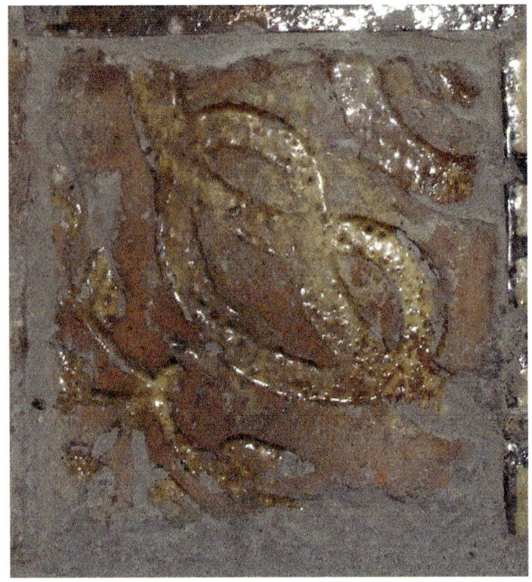

Brampton Bryan, and St Michael's, Croft.

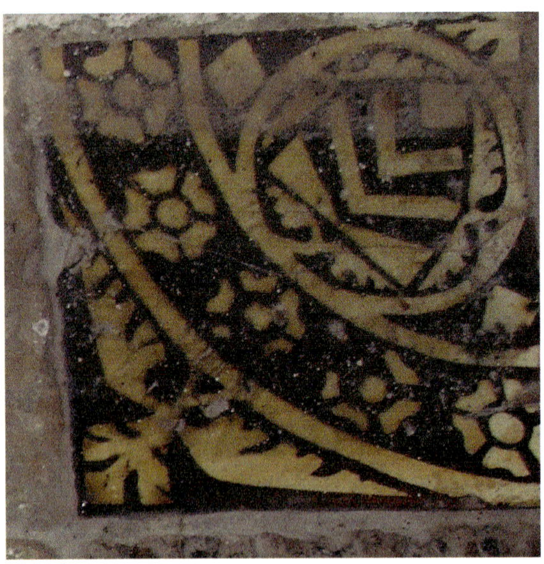

This Brampton Bryan tile matches JM Lewis 319 (opposite).

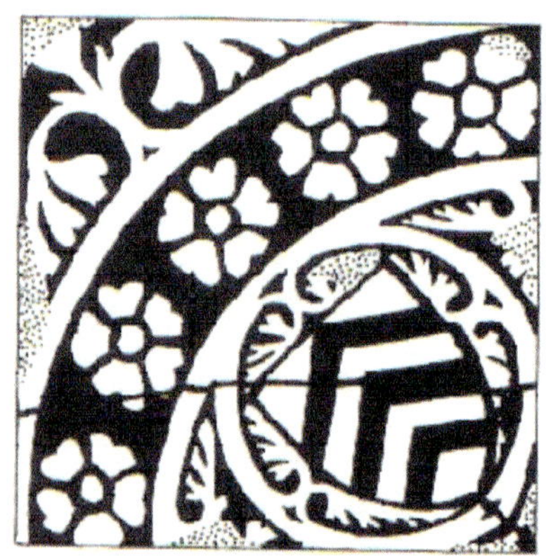

319

JM Lewis 319.

Slip-on-impression; 4 tile design; circular band of five-petalled roses surrounding four shields of arms each set diagonally within a ring of stylized foliage; *three chevrons* (Clare); outside the circle each corner contains a weeping foliage motif.

BM Design 1656 (Gloucester Cathedral, Great Malvern Priory).

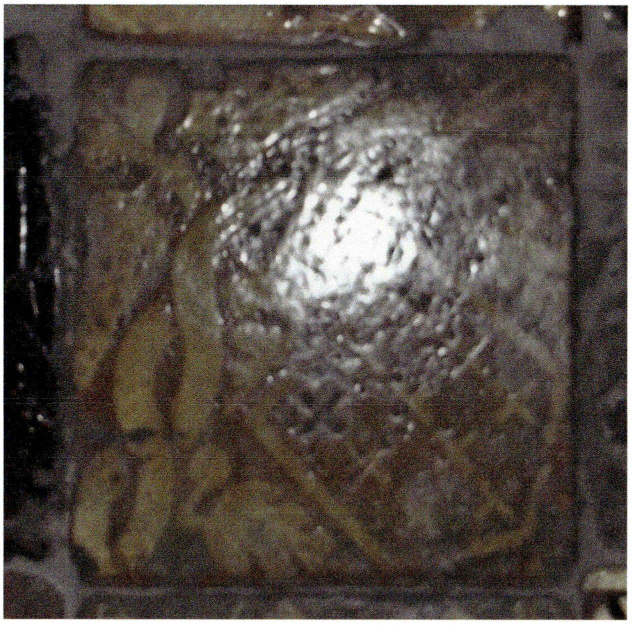

These next three designs featuring the twisted rope border around heraldic shields (10 crosses and 10 starbursts below) both have very mottled leaves in the corner which would appear to be deliberate. This fine stippling (as in BW.1) could only have been replicated with a metal (lead?) stamp.

Also at Hall Court, Much Marcle, with different shield (crossed keys - Westminster?).

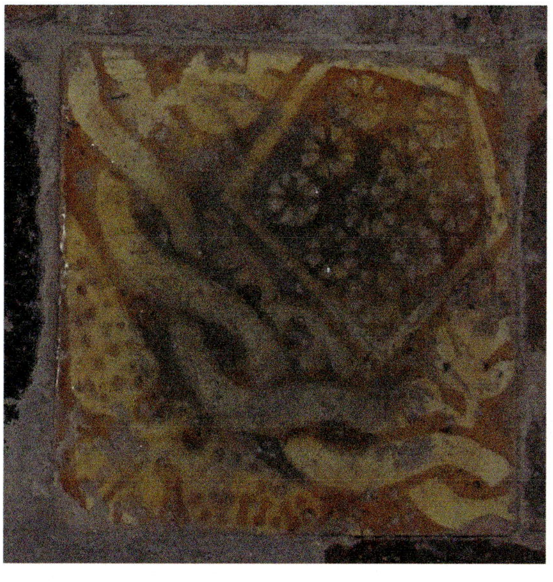

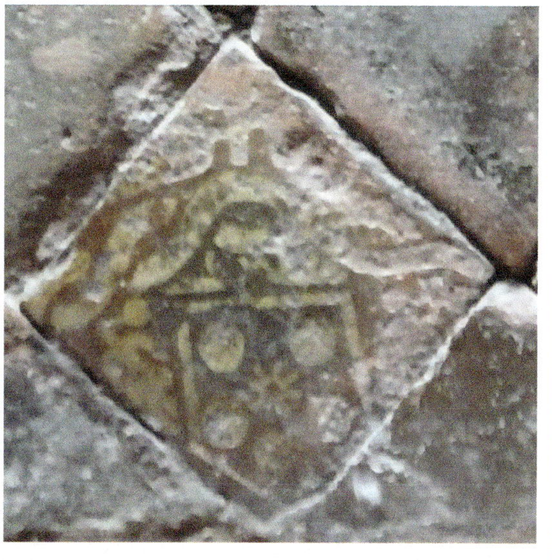

This example from St Michael's, Croft, has a different shield.

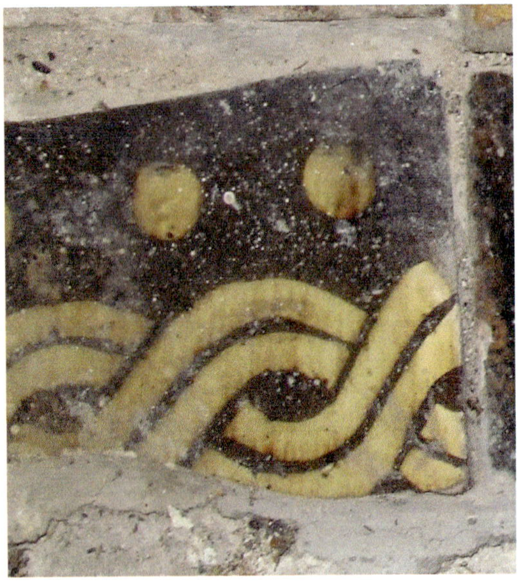

This double rope twist with dots may have been a border tile.

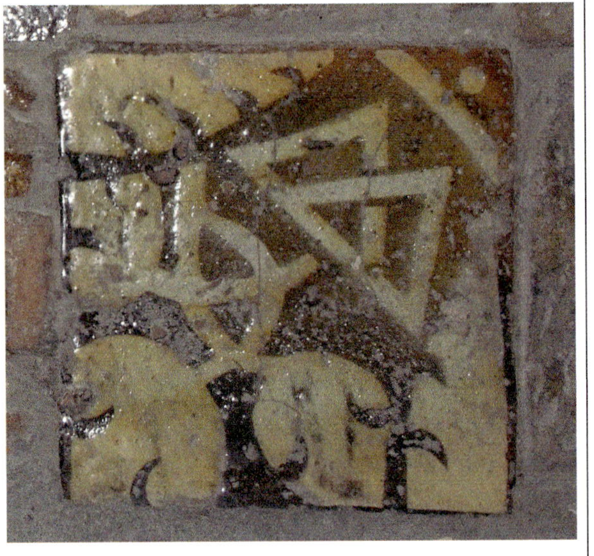

Unknown?

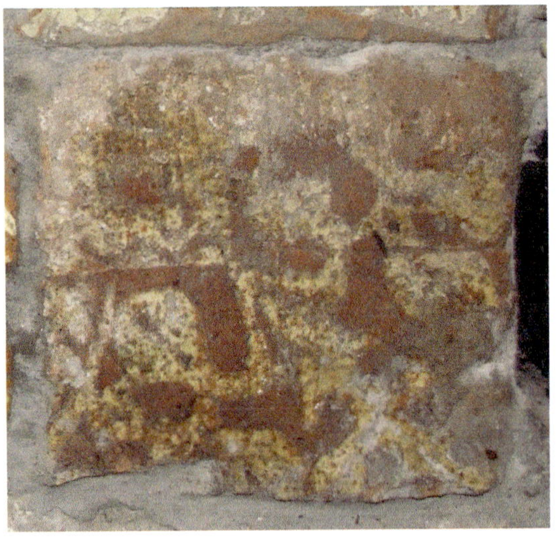

Unknown? This looks to me like an impressed design from Wigmore area.

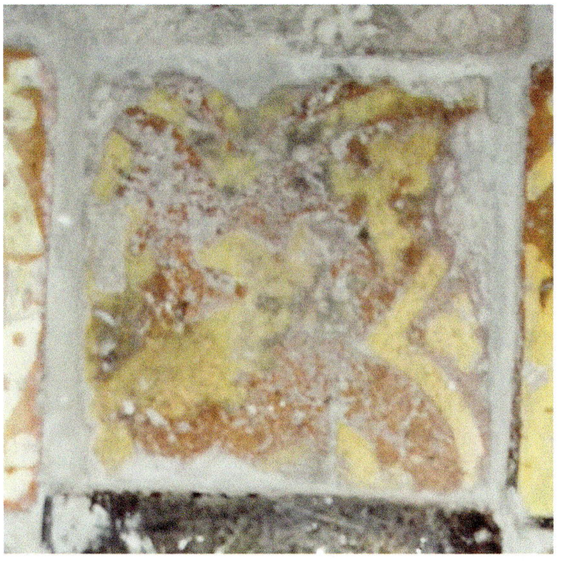	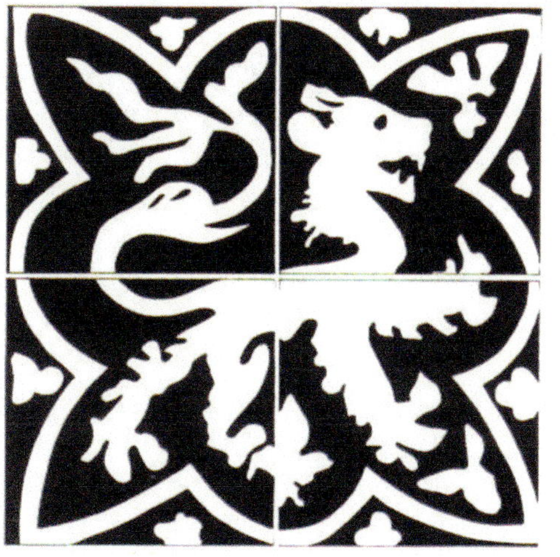 Too indistinct to recognize but possibly a cut down version of this 4-tile design from Netherwood, Thornbury.

Breinton (Lower Breinton) – Church of St Michael

A watching brief during work on the tower in February 2011 by Church and Site Archaeology Services revealed fragments of decorated medieval floortile. The one intact side measured 13cm. The fabric was very dark, brick-like, and attributed to the Malvern Chase school.

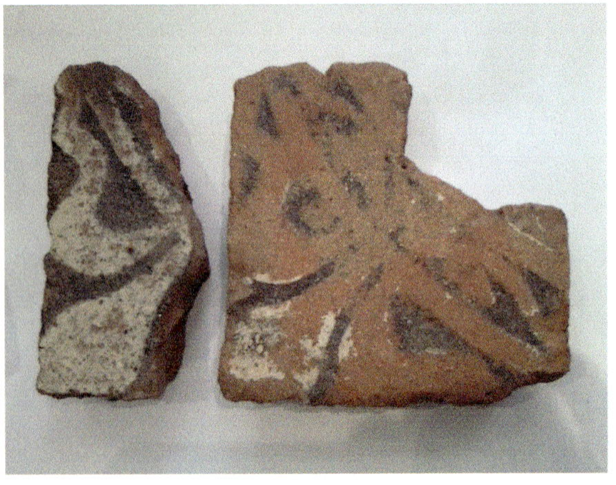

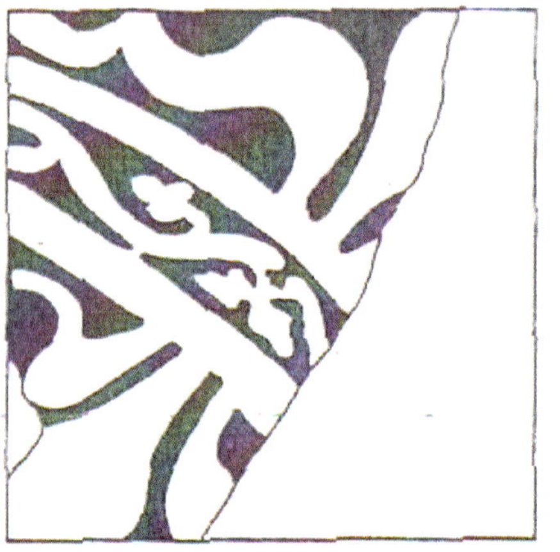

Blackfriars, Gloucester (9-tile design), and Hereford Cathedral, Booth Chapel.

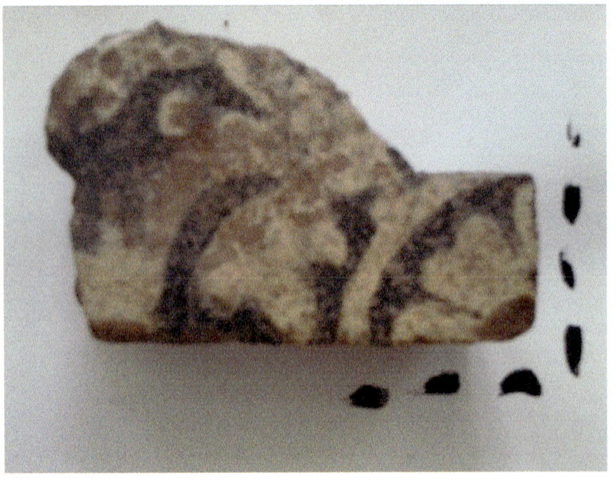

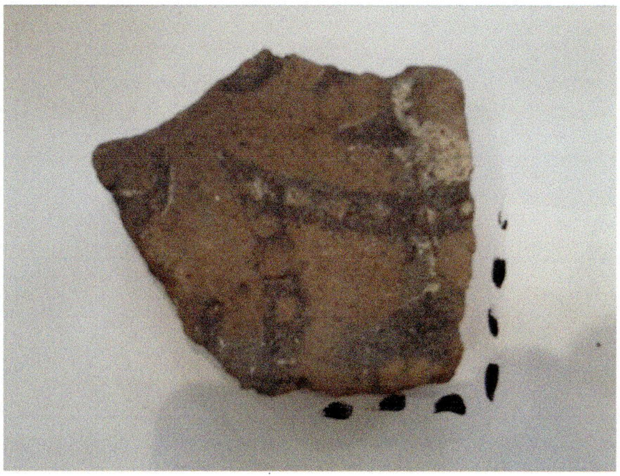

Probably 273.

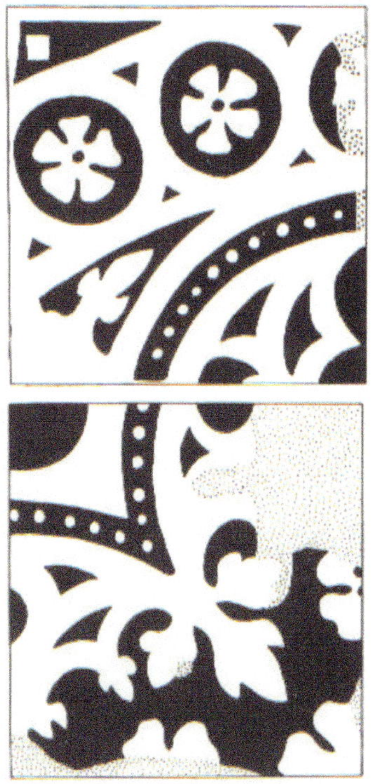

272

273

272-3 JM Lewis Group 20 (Droitwich)
16 tile design: circular band decorated with reserved roundels containing five-petalled rosettes, surrounding a quatrefoil outlined by a reserved band with pellets, and with internal tracery and foliage.
Chepstow Priory.
Hereford Cathedral, Booth Chapel.

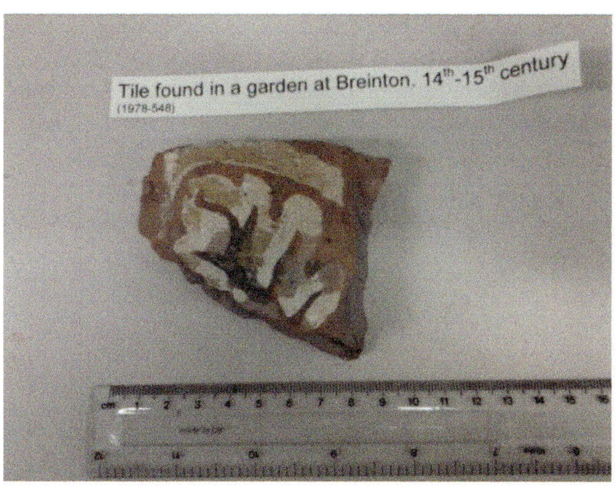

This tile from Breinton is in the collections of the Hereford Museum.

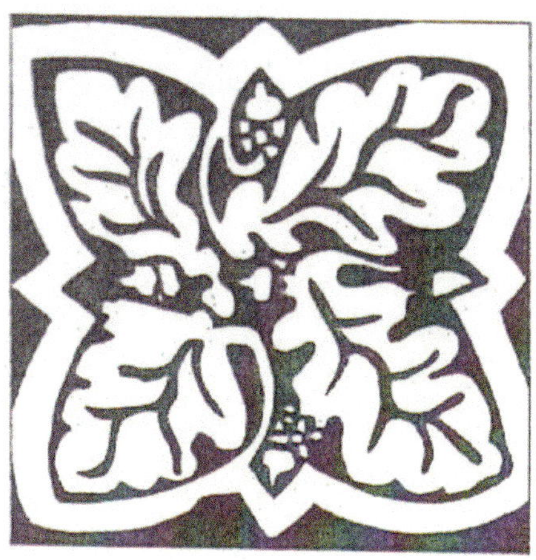

Also at Blackfriars, Gloucester.

Castle Frome – Church of St Michael and All Angels

St Michael's Church was built circa 1125, probably by a member of the de Lacey family. To the north of the church runs a sunken roadway which runs up towards the castle. The chancel arch is Norman and three original Norman doorways survive.

Under the east window in the south wall of the chancel is a plain aumbry. Medieval floortiles have been set into the back and base.

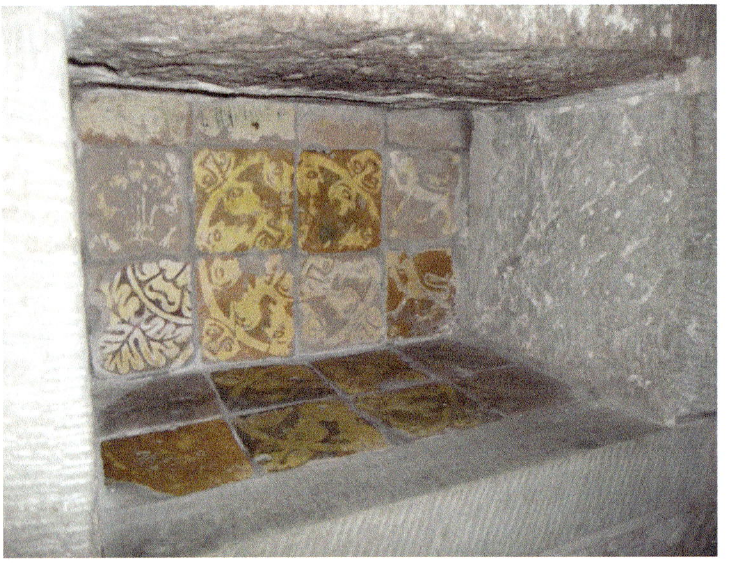

There are 16 complete tiles and four half tiles at the top of the back.

Only one tile (bottom left) can be immediately identified, and that is a Bristol Canynges Design BM 2983 JM Lewis 374 (Group 24).

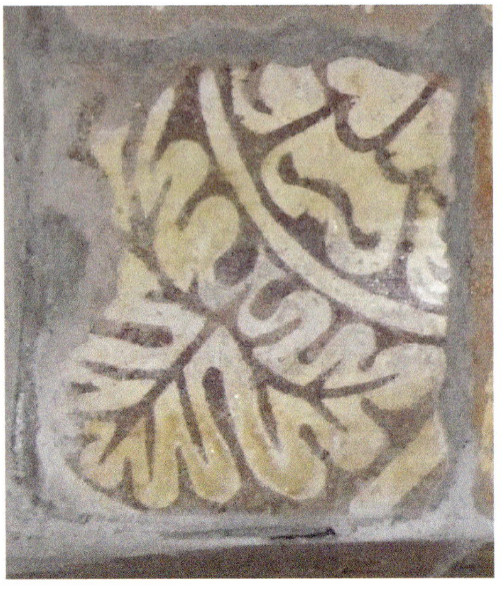	JM Lewis 371-4 16-tile design: circular bank of five-petalled rosettes enclosing a quatrefoil outlined by a band containing pointed ovals and pellets, each lobe enclosing an oak-type leaf motif; eight-petalled rose at the centre; outside the circle an oak-type leaf motif in each corner. 374 = BM 2983 Also at: Carmarthen, St David's Cathedral

All the rest of the tiles are very unusual. In particular, the two main groupings of four tiles of human mask-like faces mirrored in the base and the back of the aumbry cannot be paralled elsewhere.

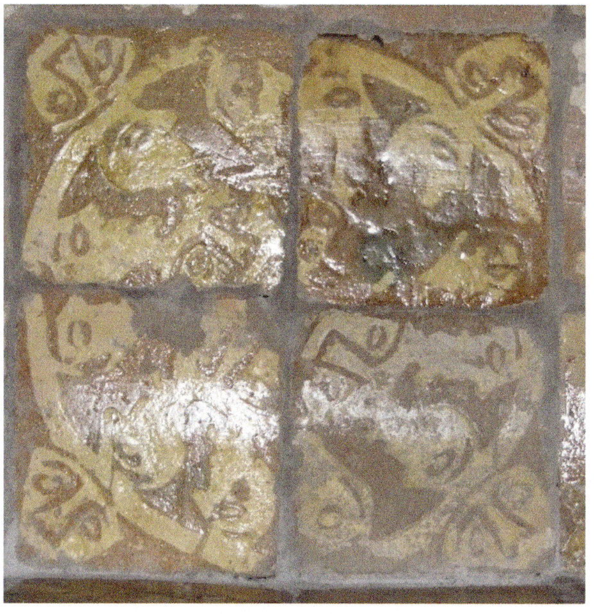

Each tile consists of three heads facing clockwise in profile with prominent nose and pointed goatee beard within a quarter circle band. At the inner and outer corner is a mask-like face, facing front. It is unusual that eight of these tiles have survived in relatively good condition, and that they were selected to be preserved at the time of the Victorian restoration in 1878, particularly when they would appear to have no obvious religious significance.

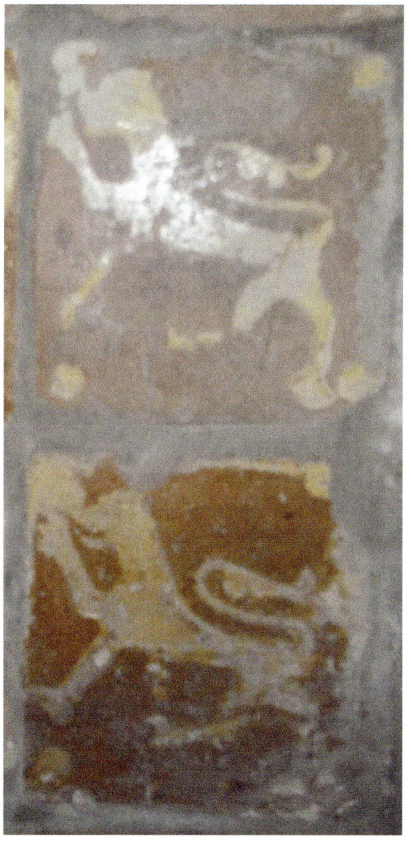

The two lions passant at the right of the back wall appear to have dots in the corners. Hilary White suggests that the stamp may have been intended for a larger tile. The lion is finely carved, suggesting an early date for its production.

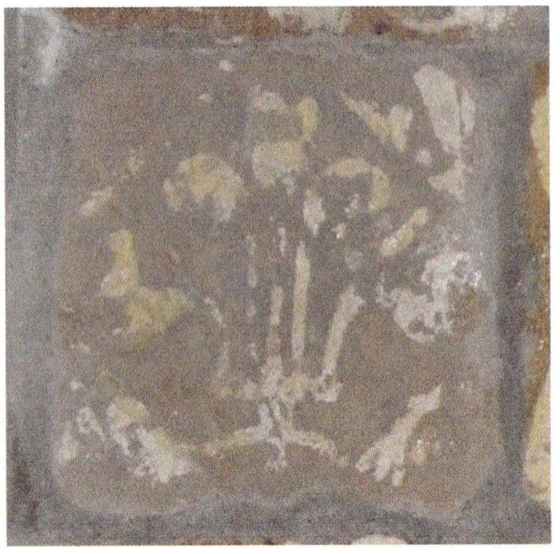	In the top left corner of the rear wall is an unusual fleur-de-lys: this is five-lobed with possibly yellow corners, and what may be animal /bird feet clutching a central 'perch', or is this simply a flaw in the design of a lozenge enclosing the fleur-de-lys?

Turning to those tiles set in the base of the recess, to the left of the central grouping of the four masks are two tiles which are very difficult to identify. In the top left hand corner is a worn tile of tracery / foliate design.

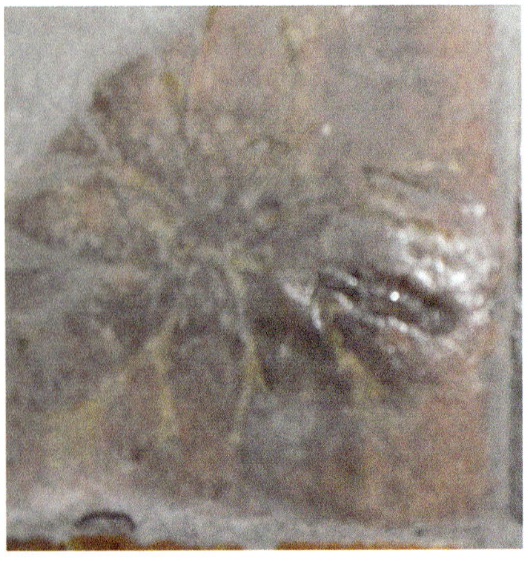	This is similar to a design found in All Saints, Hereford. 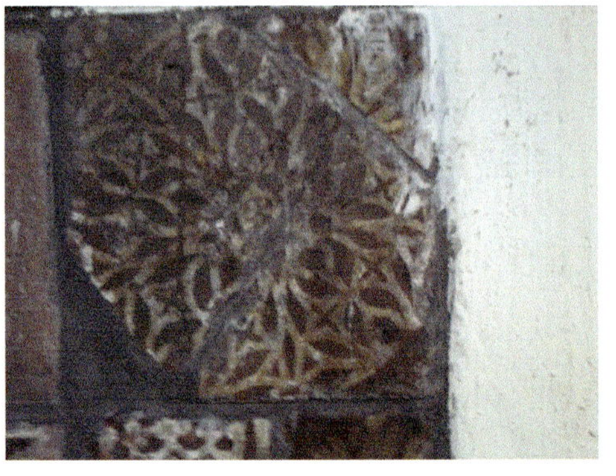
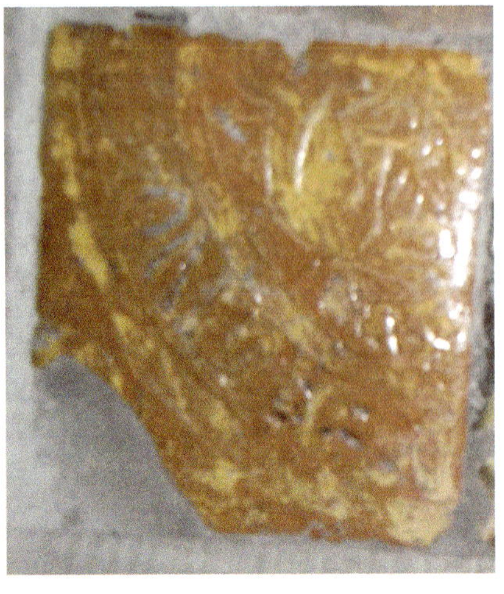	In the bottom left corner is a tile with a very bright orange glaze remaining. It is clearly one of a 4-tile group and possibly contains text within the outer band. An indistinct delicate foliage design fills the centre.

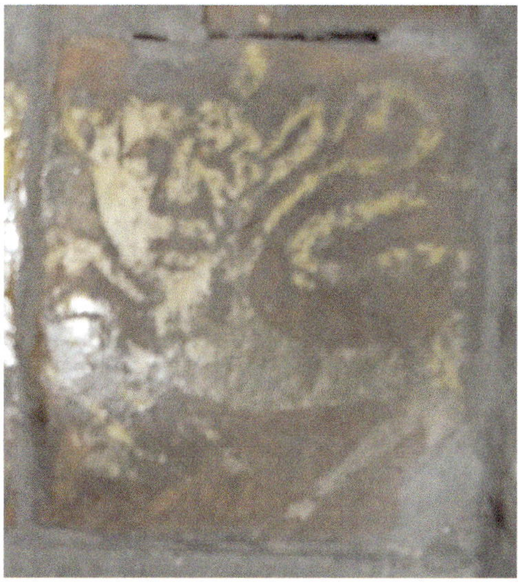	To the right of the mask-like faces is a very unusual design of a beast with a large human face, facing front, with large ears and a long curly tail. Again the stamp seems slightly too big for the tile, but is finely drawn, suggesting an early date.
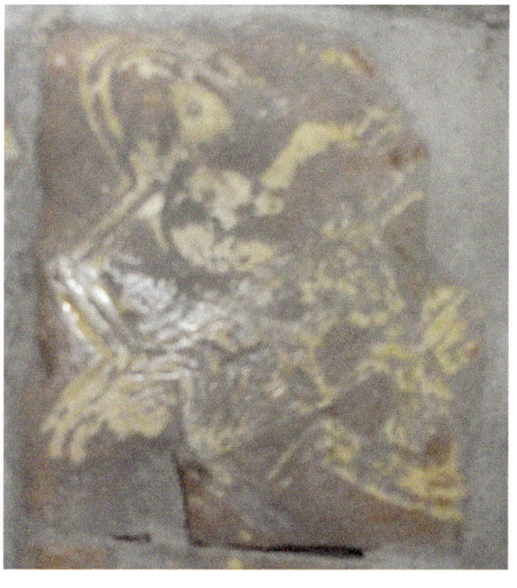	Behind this is a damaged tile with parallels from Blackfriars, Gloucester; Hereford Cathedral's Booth Chapel; and also Limebrook Nunnery. The multiple-toed beast has the face sometimes of a cat/lion, or of a nun. The triple border is unusual; a similar tile with double banding is known from Evesham Abbey. 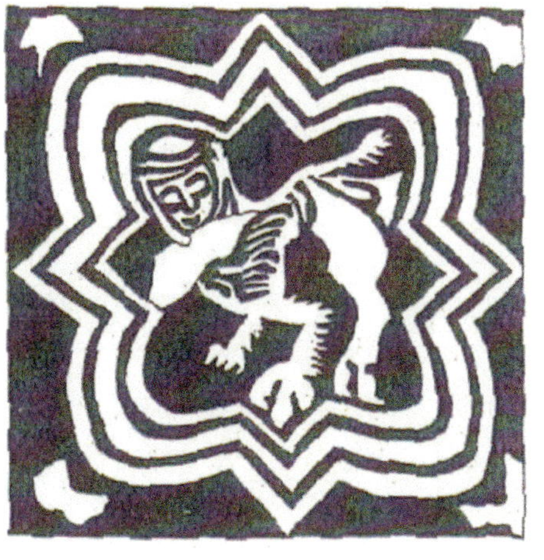 Blackfriars, Gloucester

Colwall – Church of St James

In the north wall of the north aisle there is a round sliptile depicting a labourer digging. This is thought to be one of a group of tiles featuring the Labours of the Months and likely to date to the late thirteenth century.

(All photos from Colwall are courtesy of Laurence Keen.)

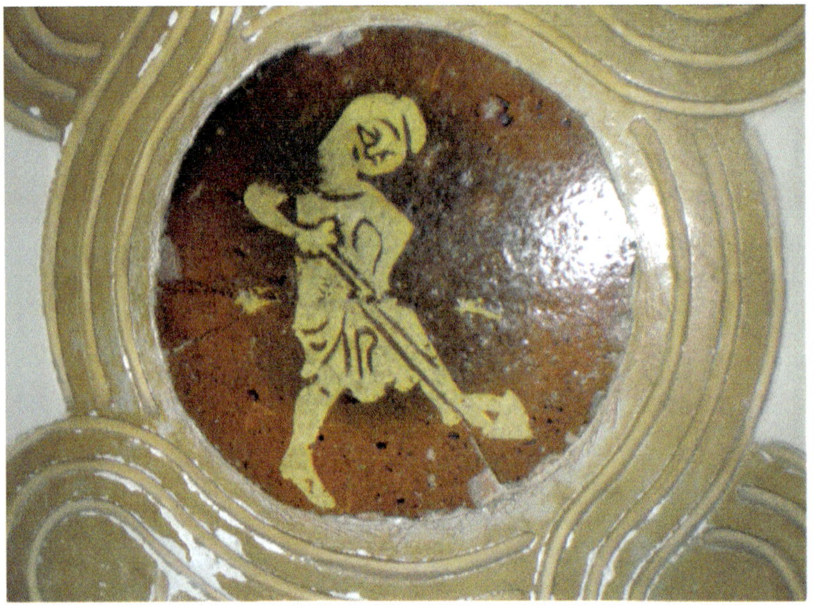

On the sill of the second north window are fragments of tiles. Where drawings of the designs are available I have used these, as the Colwall tiles are difficult to photograph clearly.

	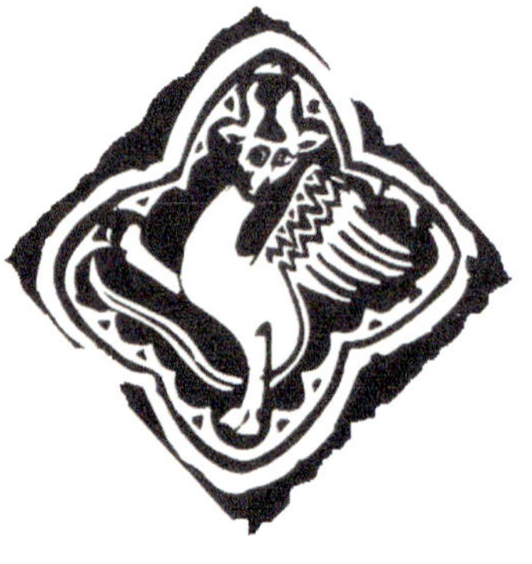
This drawing is of a tile from Blackfriars, Gloucester. This design is also found at Hall Court, Much Marcle, and in the Booth Chapel at Hereford Cathedral.	*After Eames* BM 1389. Also at Dore Abbey, Eign Gate, Hereford, and Blackfriars, Gloucester.

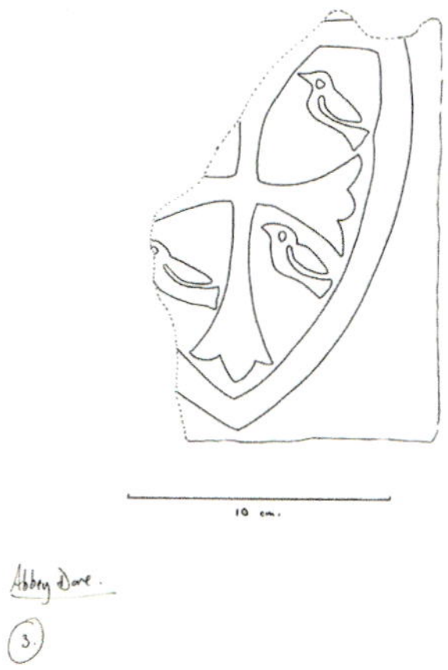 Also at Dore Abbey: *"a cross paty between four martlets"* C14 Westminster Abbey – mother house of Gloucester Abbey.	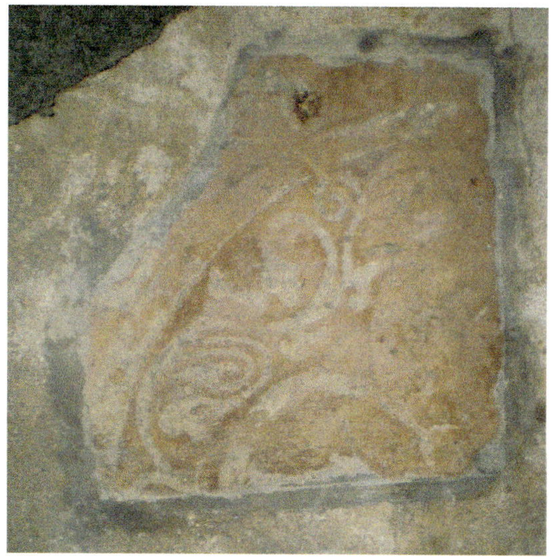 Unidentified.

Croft – Church of St Michael and All Angels

The church dates from the mid fourteenth century and is inevitably closely linked with the castle nearby. Between 400-500 floortiles, taken from the chapel of Croft Castle which was demolished in the seventeenth century, have been relaid randomly in the nave, chancel, and in the step risers to the chancel.

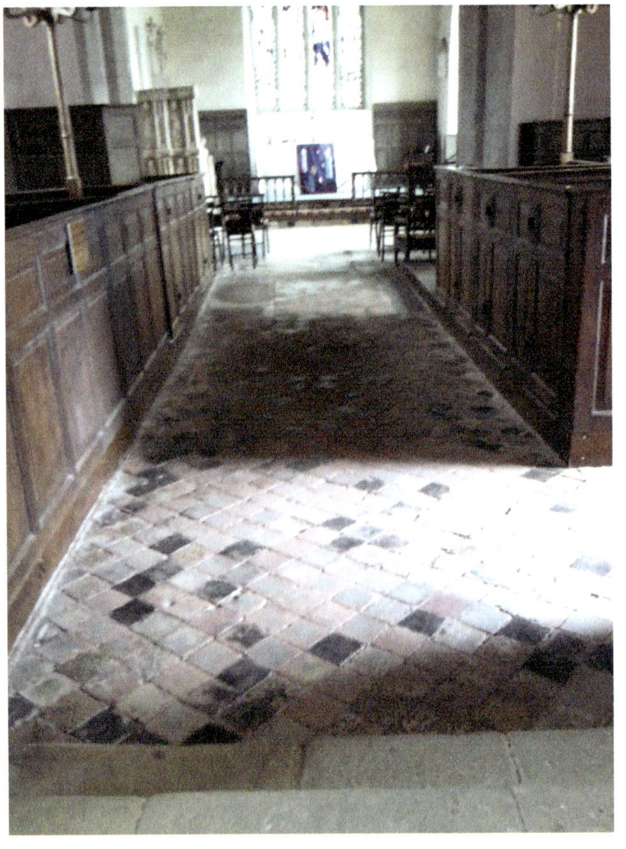

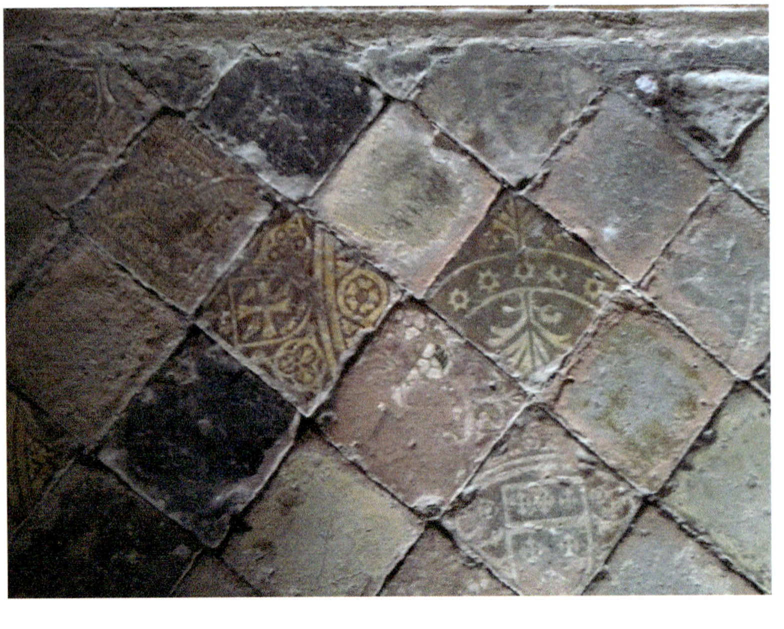

Alan Vince believed that the tiles at Croft, Stretton Sugwas, and All Saints, Hereford were supplied from Great Malvern or a daughter tilery between 1450-1460. One tile (324) bears the date of 1456.

The similarities with the Stretton Sugwas tiles are overwhelming. Using the JM Lewis tile numbers, the following tiles are to be found both at St Michael's, Croft, and at Stretton Sugwas:

304, 307, 308 (with a different centre) 310, 312, 318, 322, 324, 332, 333, 337, 338, and 342.

Croft alone has the following tiles (not at Stretton Sugwas):

315 (shield containing instruments of the passion)
317 (4-tile circle containing 4 crowned Ms)
323 (4-tile square (pas xpi)

Stretton Sugwas alone has the following tiles (not at Croft):

311 (the crowned M),
320 (the Beauchamp arms)
331 (turrets and pinnacles)
334 (with instruments of the passion)
339 (design with circles and two birds at the corner)
341 (intricate tracery design)
343 (the chained swan of ?The Mohuns ? or the Stafford family)
345 (portcullis with text in circular band)

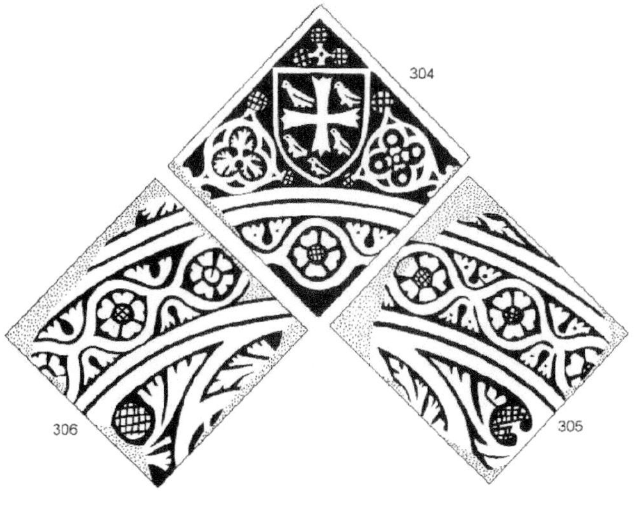

JM Lewis 304-6 Slip on impression: 16-tile design: circle of linked five-petalled roses flanked by double lines enclosing a quatrefoil containing leaf designs; outside the circle in each corner a shield of arms; *a cross patonce between five martlets* (Westminster Abbey, the mother house of Gloucester Abbey), flanked by two traceried roundels containing a trefoil and a quatrefoil respectively.

304 = BM Design 2897
305 = BM Design 2898
306 = BM Design 2899
(Great Malvern Priory).

Also at Stretton Sugwas, Raglan Castle, and Monmouth.

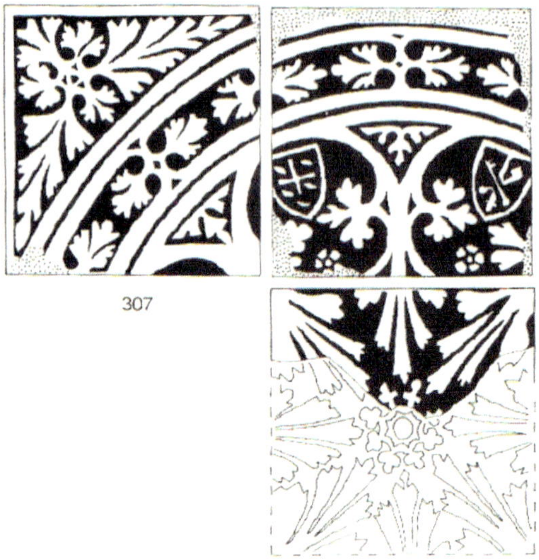

307 308 309

JM Lewis 307-9 Slip on impression: 9-tile design; circle of double lines flanking a frieze of opposed three-leaf motifs: within the circle leaf tracery forming eight compartments containing shields of arms of two designs arranged alternately: *a cross patonce between five martlets* (Westminster Abbey) and *cross keys with sword in pale* (Gloucester Abbey); in the centre an octagonal motif of linked trefoils; outside the circle each corner has a framed leaf motif.

307 = BM Design 2859 (Gt Malvern)
308 = BM Design 2860
(Gloucester Cathedral; Great Malvern)

309 = BM Design 2565 (Great Malvern)

336

At Stretton Sugwas this group has a different central tile 336 – BM 2435.

Also at Monmouth, Abergavenny, and Llangattock-nigh-Usk.

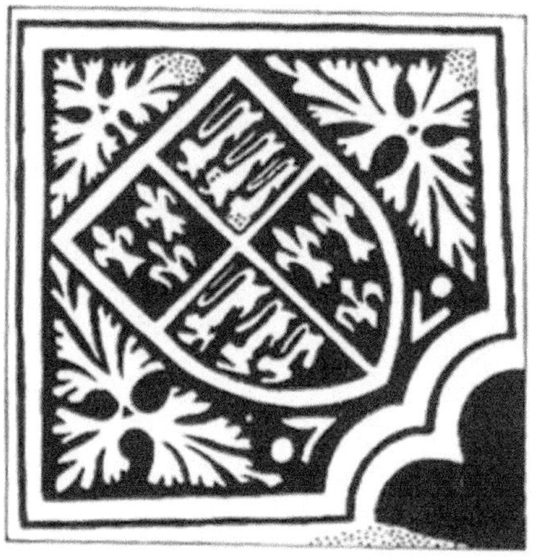

310

JM Lewis 310 – 15 Slip on impression:
4-tile or larger arrangement consisting of a grille of reserved quatrefoils with the arms of England at the corners.

310 Slip on impression:
Shield of arms set diagonally : *quarterly 1st & 4th three fleur-de-lys 2nd & 3rd three lions passant guardant* (royal arms of England, post 1405) within plain double border with quarter quatrefoil in bottom corner; the corners within the border are filled with foliage decoration.

310 = BM Design 1518
(Gloucester Cathedral, Great Malvern Priory, Lenton kiln site)

Also at Stretton Sugwas, Abergavenny, Monmouth and Raglan.

312

JM Lews 312 Slip on impression:
Centre tile; crowned monogram *ihc* within double circle; quarter quatrefoils in each corner.

312 = BM Design 1418
(Great Malvern Priory)

Also at Stretton Sugwas, and Monmouth.

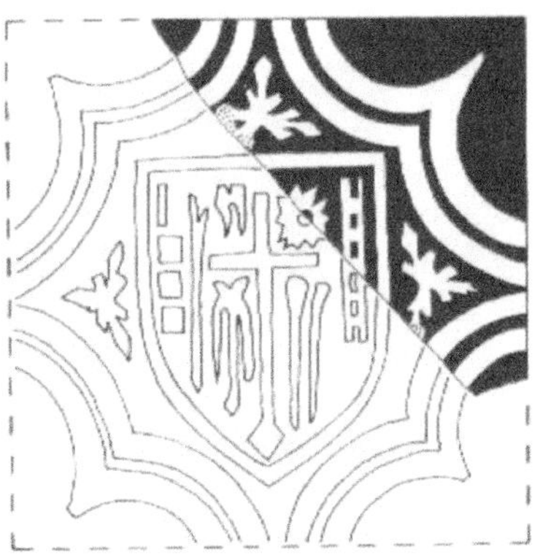

315

JM Lewis 315 slip on impression:
Centre tile: shield of arms set squarely in centre containing the Instruments of the Passion, quarter quatrefoils in each corner.

315 = BM Design 1693
(Great Malvern Priory)

Also at Monmouth.

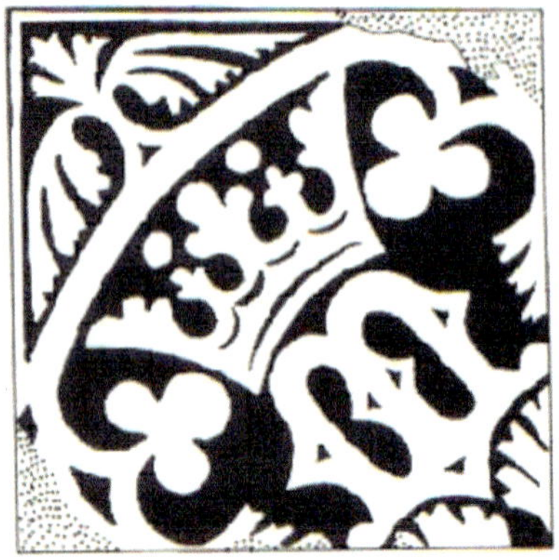

317

JM Lewis 317 Slip on impression:
4-tile design; circle containing four crowned Ms separated by tracery and flanked by trefoils; at the centre a quatrefoil containing foliage decoration; outside the circle each corner contains a weeping foliage motif.

317=BM 1425 (Great Malvern Priory, Lenton Priory kiln site)

Also at Monmouth and Raglan.

318A

JM Lewis 318 Slip on impression:
4-tile design; a barbed quatrefoil frame with repeating inscription: *fiat voluntas dei* (May the Will of God be done), surrounding four shields of arms; *three lions passant guardant* (royal arms of England, pre-1340); outside the frame each corner has a pair of birds flanking a central stem.

318 = BM Design 1480
(Gloucester Cathedral, Great Malvern Priory)

Also at Stretton Sugwas, Raglan, Llangattock-nigh-Usk, and Monmouth.

322

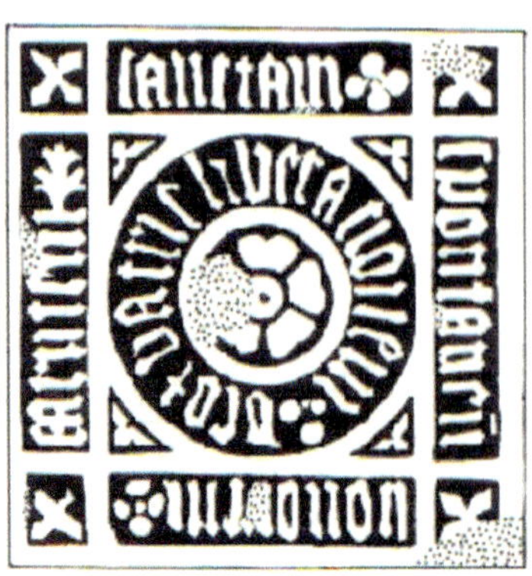

JM Lewis 322 Slip on impression:
Square subdivided into a central square by narrow rectangular borders; the centre has a central five petalled rose surrounded by a circular band; the borders and central ring are filled by a continuous black-letter inscription reading: *mentem sanctam, spontaneu(m) honorem deo, et patrie liberacionem* (A holy mind, honour freely given to God, liberty to the country).
These words refer to St Agatha and occur in her *Life* (Bollandius & Herschinjus 1863, 640A). They were used in the Middle Ages as a charm against fire.

322 = BM Design 1429
(Great Malvern Priory, Lenton Priory kiln site)
Also at Stretton Sugwas, Monmouth, and Raglan.

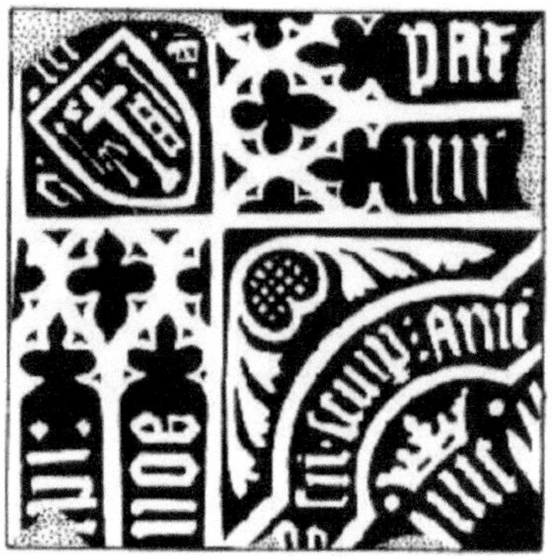

323

JM Lewis 323 Slip on impression:
4-tile design; subdivided square, with central barbed quatrefoil surrounded by a double border in the form of two-light windows; a shield of arms bearing the Instruments of the Passion in each corner; on the borders of the tile and the quatrefoil a black-letter inscription repeated four times: *pas xpi; int(er)' nos sit sempl(er): Ame'* (May the peace of Christ be amongst us always. Amen); within the quatrefoil *ihc* crowned.

323 = BM Design 1469
(Great Malvern Priory)

Also at Monmouth, and Raglan.

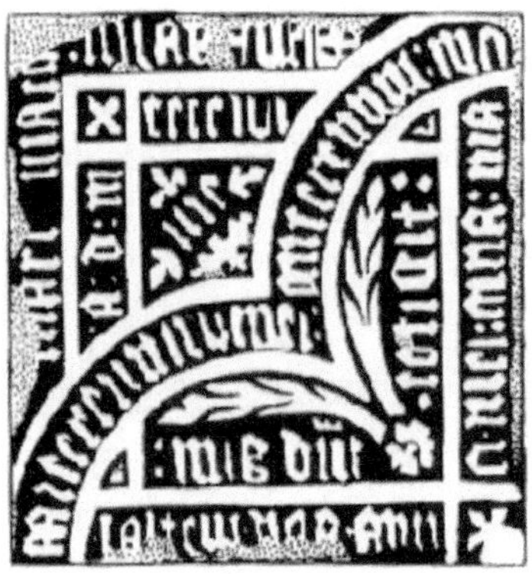

324

JM Lewis 324 Slip on impression:
4-tile design; ribbon quatrefoil with double border outside and equal-arm cross inside; all the spaces are filled with a continuous inscription repeated four times: *marc mathe:lucas:joh/ a:d:mccccIvi/ miseremini:mei:miseremini:mei/ saltem.vos.amici.mei:quia:ma/nus dni.tetigit:/me* (Mark, Matthew, Luke, John. A:D: 1456. Have pity upon me, have pity upon me at least ye my friends, for the hand of the Lord hath touched me). The words include a passage from Job xix,21 used in the Office of the Dead.

324 = BM Design 1468
(Great Malvern Priory, Baynham Abbey)
Also at Stretton Sugwas, Llangattock-nigh-Usk, Llanthony Priory, Monmouth, and Raglan.

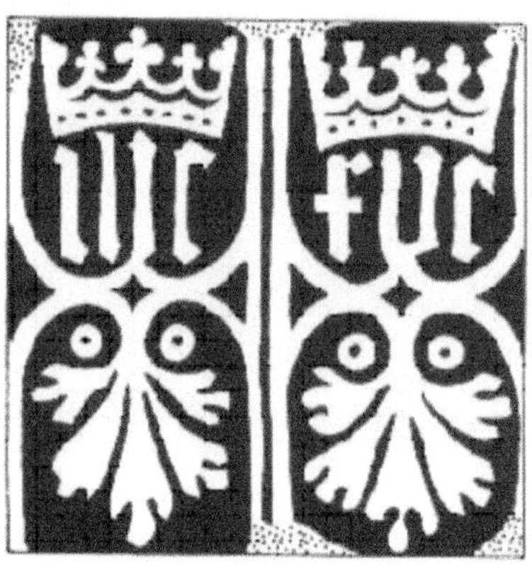

332

JM Lewis 332 Slip on impression:
Two-light opening with pendant leaf motifs, and with crowned monograms above: *ihc* and *xpc* (Jesus/Christ).

332 = BM Design 1420
(Great Malvern Priory, Lenton Priory kiln site)

Also at Stretton Sugwas, Hereford All Saints, Abergavenny, and Monmouth.

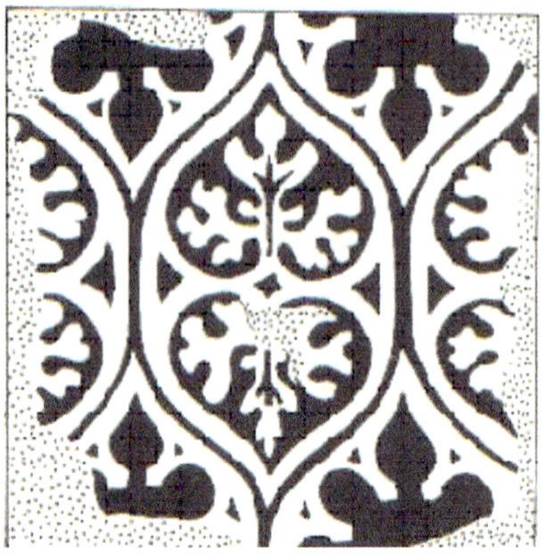 333	JM Lewis 333 Slip on impression: Two-light tracery. 333 = BM Design 2575 (Great Malvern Priory, Lenton Priory kiln site) Also at Stretton Sugwas, and Monmouth.
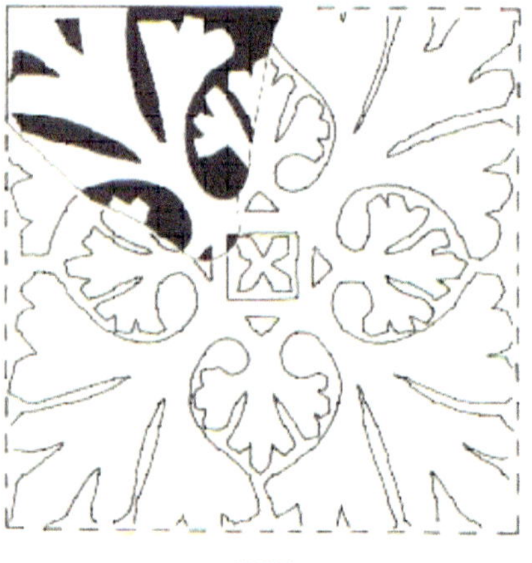 337	JM Lewis 337 Slip on impression: Foliage decoration arranged in the form of a cross set diagonally with smaller areas of decoration between the arms. 337 = BM Design 2533 (Great Malvern Priory, Lenton Priory kiln site) Also at Stretton Sugwas and Raglan.
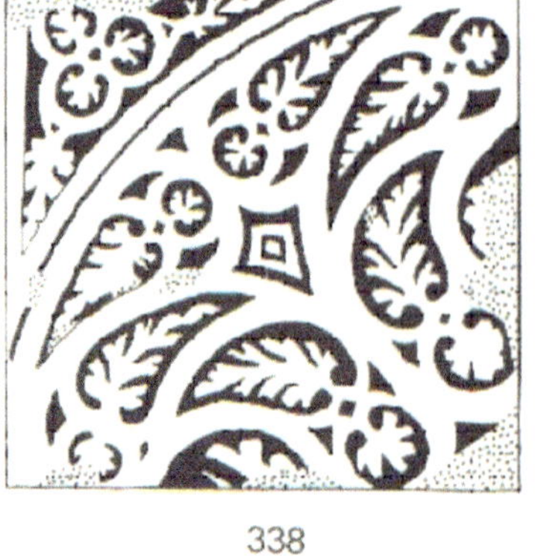 338	JM Lewis 338 Slip on impression: 4-tile design; circle with elaborate arrangement of foliage designs within reserved mouchettes; outside the circle each corner is occupied by a trefoil and leaf motif. 338 = BM Design 2689 (Great Malvern Priory, Lenton Priory kiln site) Also at Stretton Sugwas, Monmouth, and Raglan.

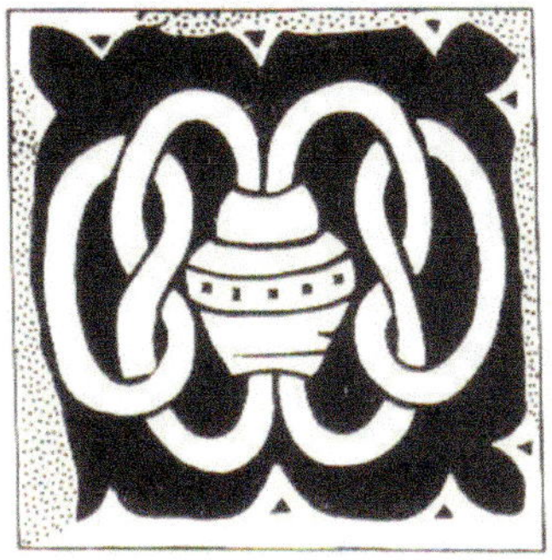

342

JM Lewis 342 Slip on impression:
Within a square cusped frame a Stafford knot surrounding the nave of a wheel. The badge of the Stafford family, probably for Humphrey Stafford, Duke of Buckingham, d.1483

342 = BM Design 1739
(Great Malvern Priory)

Also at Stretton Sugwas, Llangattock-nigh-Usk, Llantilio Pertholey, and Monmouth.

The following tiles at Croft are also found at Brampton Bryan:

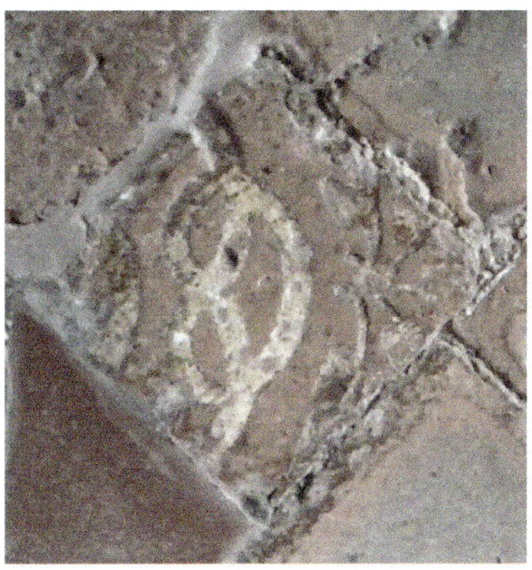

St Michael's, Croft.

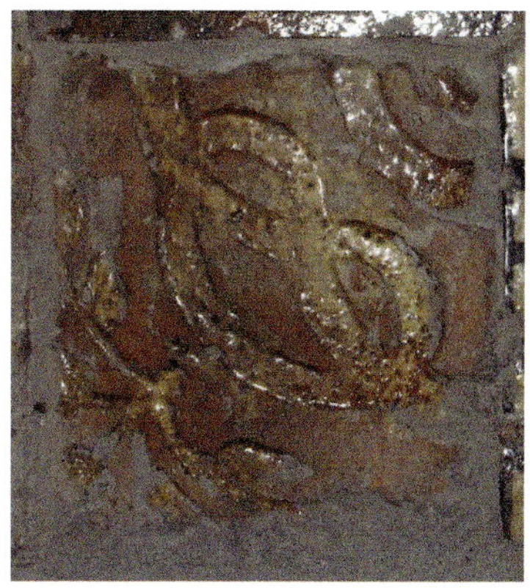

c/f from Brampton Bryan.

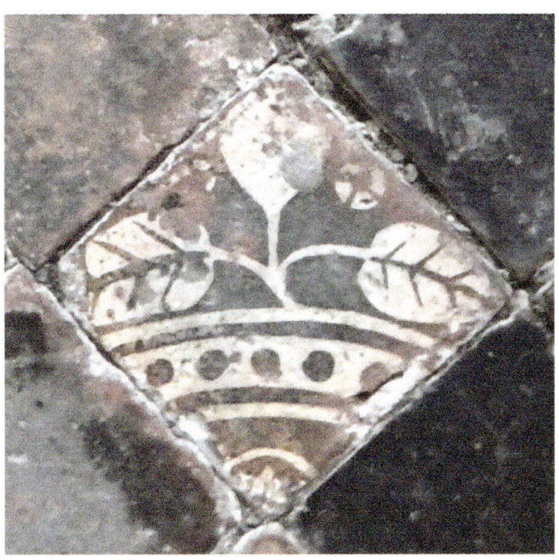

Also at Brampton Bryan.

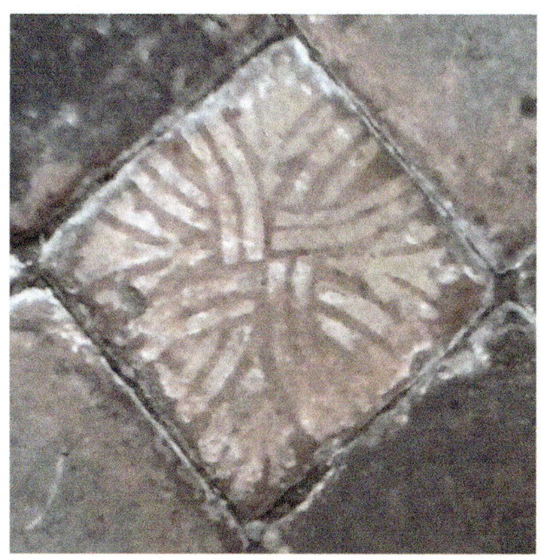

Also at Leominster Old Priory, and Brampton Bryan.

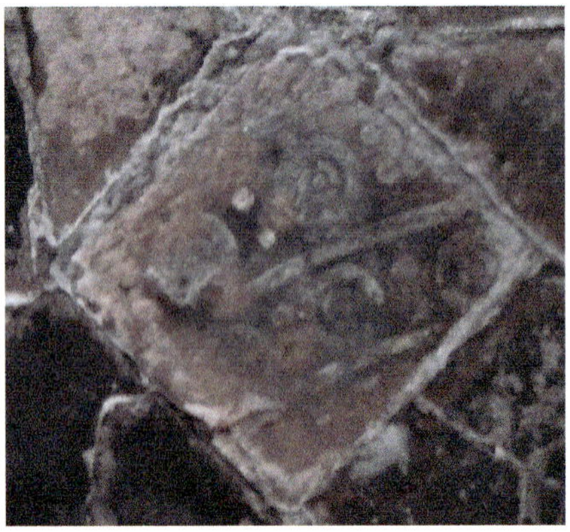

Also at Leominster Old Priory (BW 36), and Brampton Bryan.

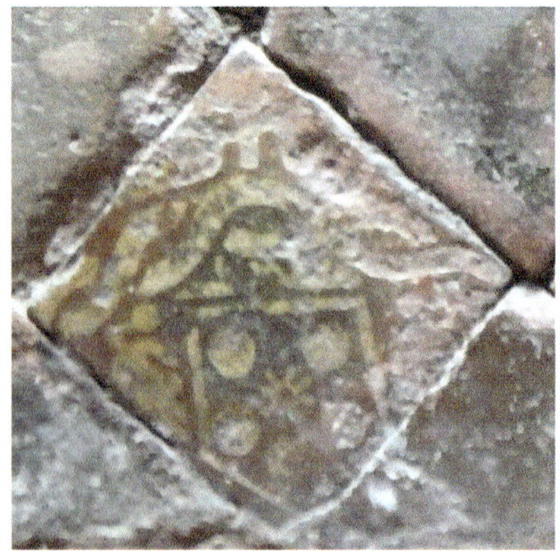

Similar at Brampton Bryan and Much Marcle – cable twist but with different heraldic shield.

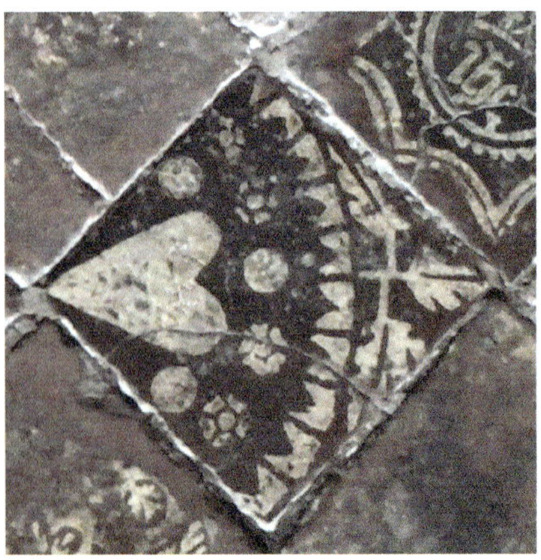

Also at Brampton Bryan and Wigmore Abbey.

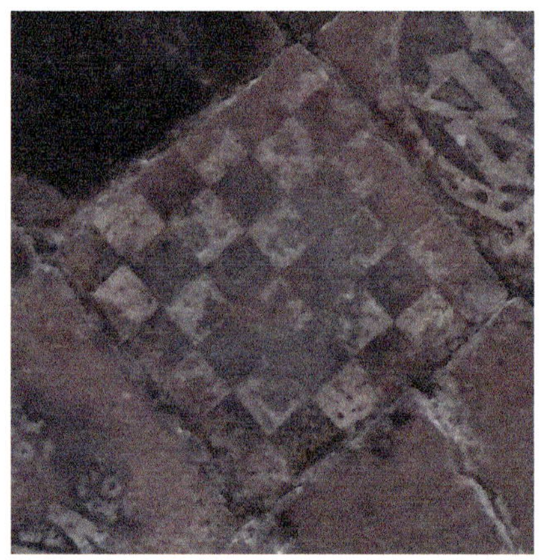

Also at Leominster Old Priory, and Brampton Bryan.

Also at Brampton Bryan (ihc).

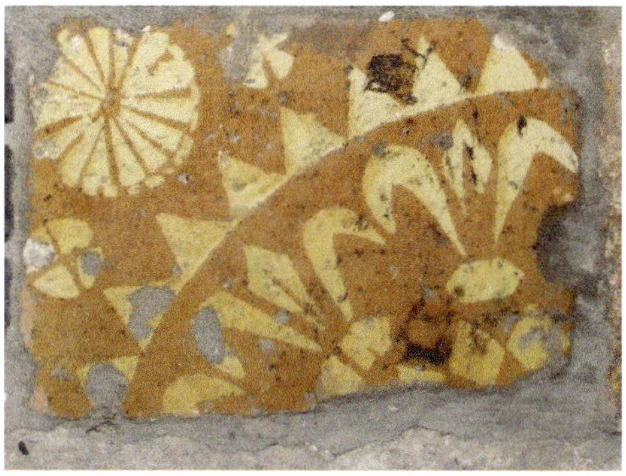 This tile is from Brampton Bryan; the photo from Croft is indistinct.	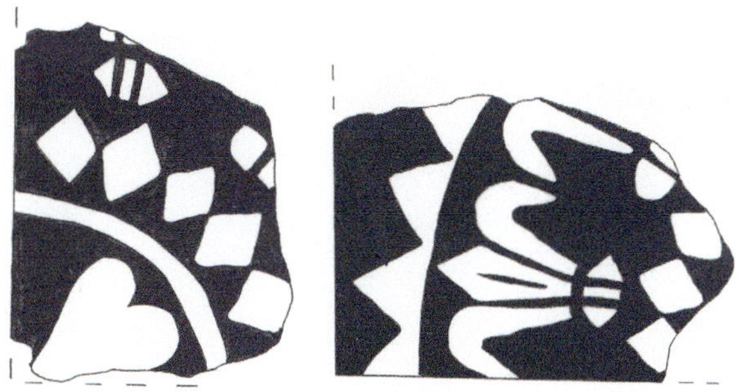 This drawing is of the same design from LOP BW 5-8.

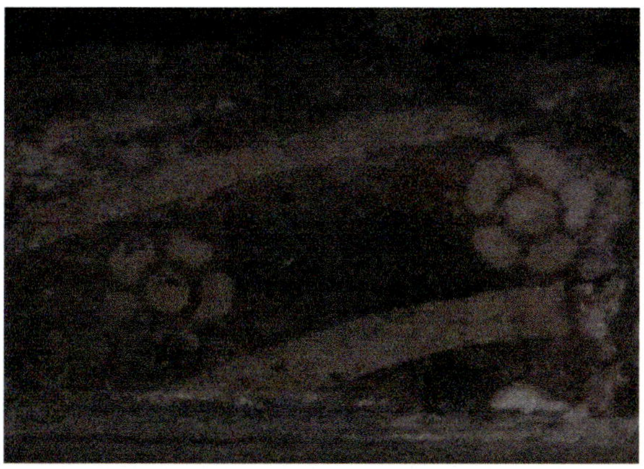	Very like Leominster Old Priory BW 35. 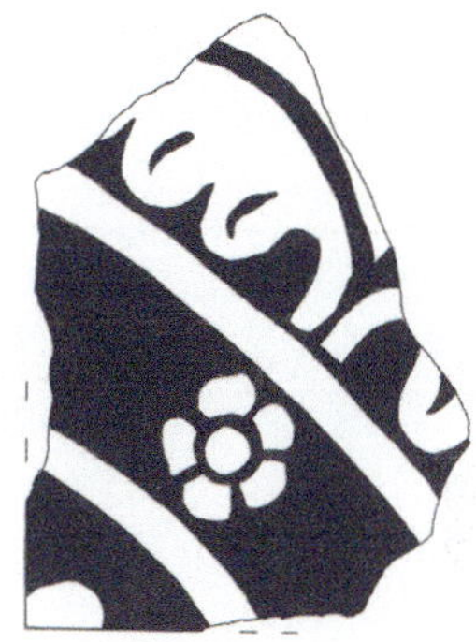

The following tiles can be identified from other sources:

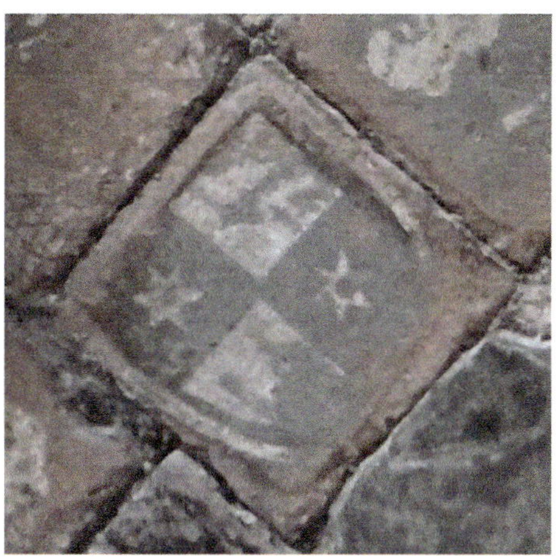	Identified by Vince & Wilmott as "Tewkesbury No.51" Robert of Oxford was a brother in law of Mortimer of Wigmore though he held little land in the Marches. *"Quarterly in the first quarter a mullet"* 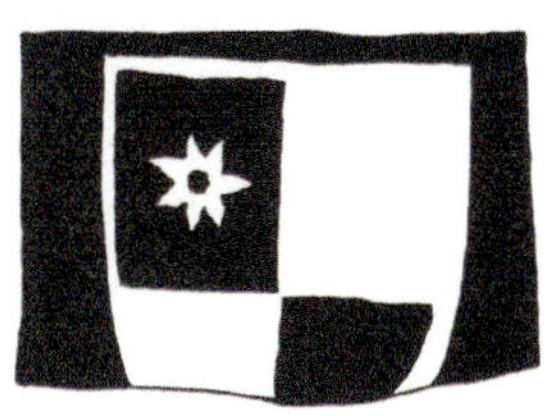 51 Also at Leominster Old Priory, and Bredon.

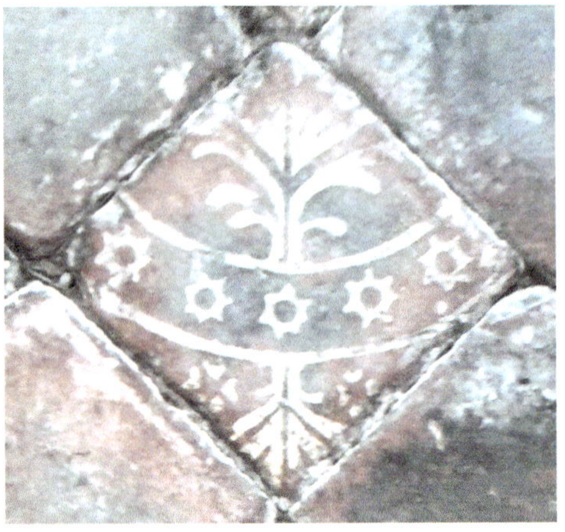

After Eames BM 2800
Lilleshall Abbey, West Midlands 13/14 century.

The following have no published replica designs that I can find:

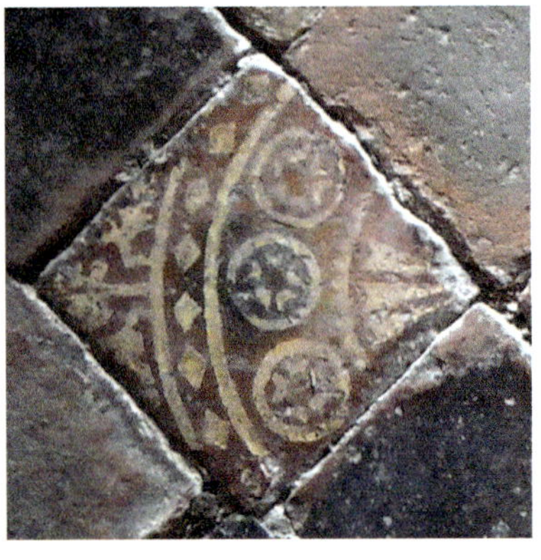

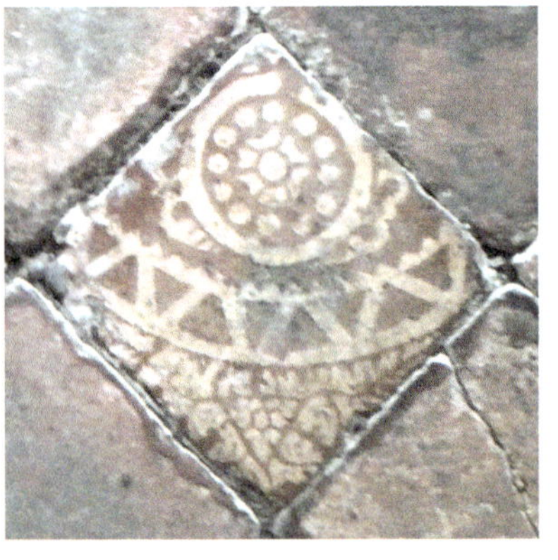

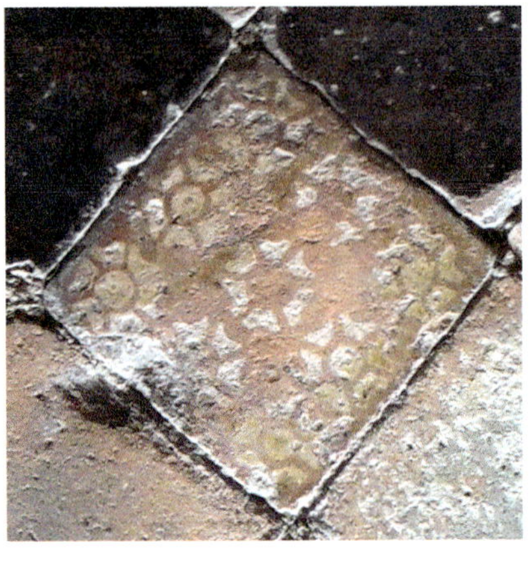

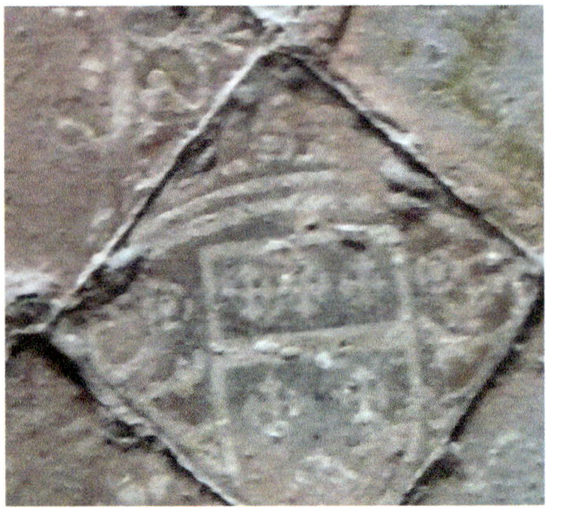

Shield of Beauchamp family.

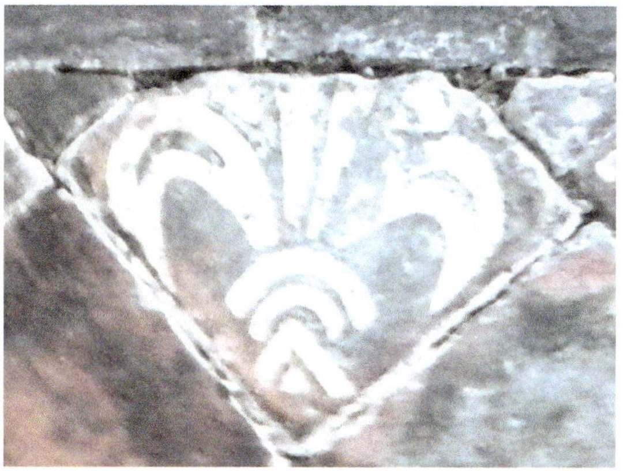

Dilwyn – St Mary's Church

Pevsner records that 'there are a number of fourteenth century tiles at the west end of the south aisle.' There is in fact a random jigsaw of small fragments of floortile remaining, but a watching brief by Church and Site Archaeology during 2005 on work in the tower uncovered slightly larger fragments of very similar tiles.

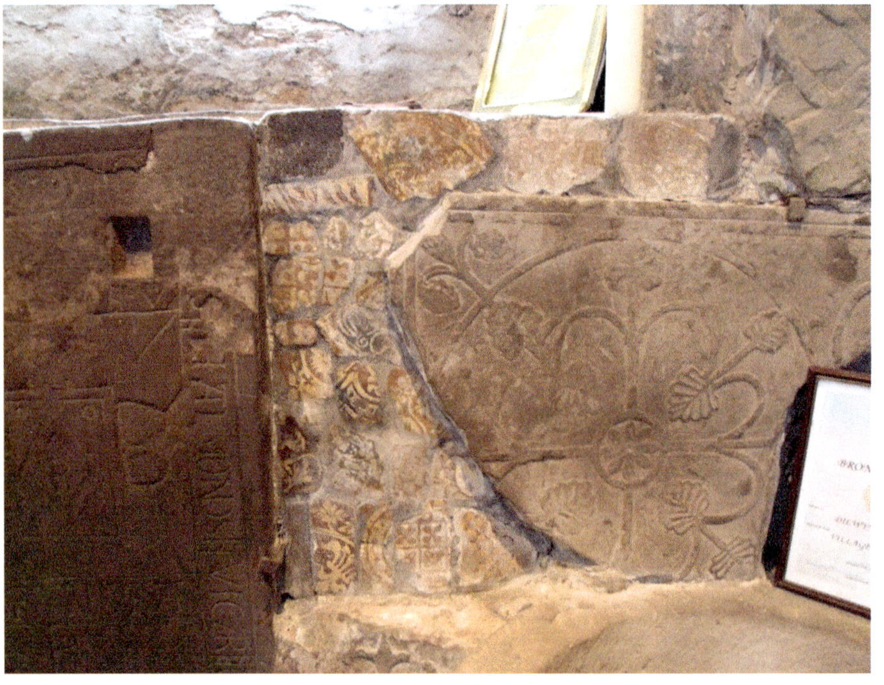

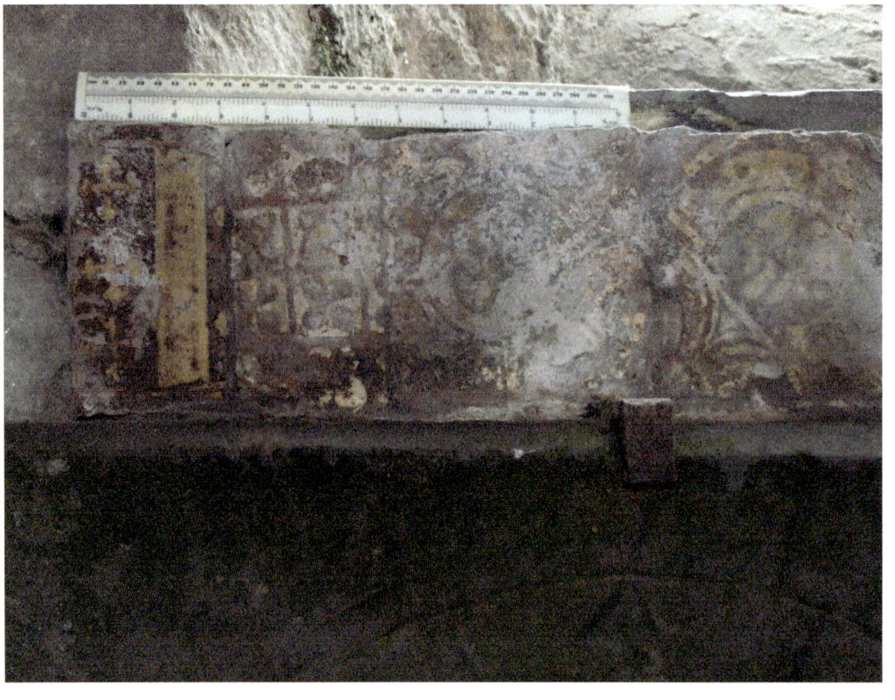

Most of the tiles appear to have formed a square 15.5 – 16cm, although a few examples are smaller, between 12.5 – 13cm. The thickness ranges from 2-2.5cm.

The outstanding feature of these designs is their close similarities to tiles found at Blackfriars, Gloucester, and recorded by Laurence Keen. I am grateful to him for allowing use of his drawings.

I have shown the Dilwyn tiles in the left hand column to compare with those from Blackfriars, Gloucester, in the right hand column. The drawings of the Dilwyn tiles are by the author.

Two of the Dilwyn tiles exhibit the same flaws in the stamps as tiles from Blackfriars, suggesting a very close link between these two sites.

DILWYN	BLACKFRIARS, GLOUCESTER
62 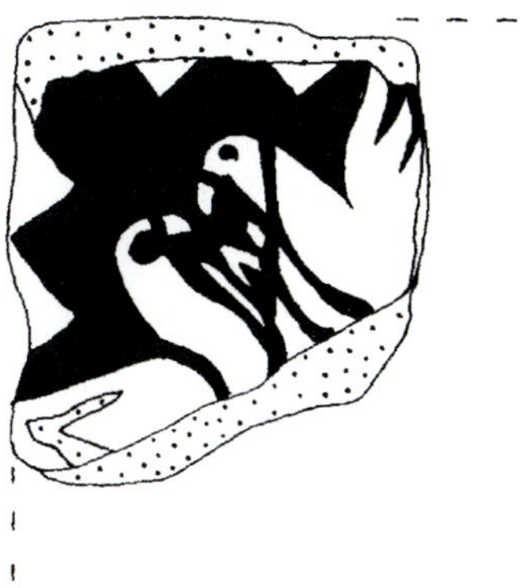	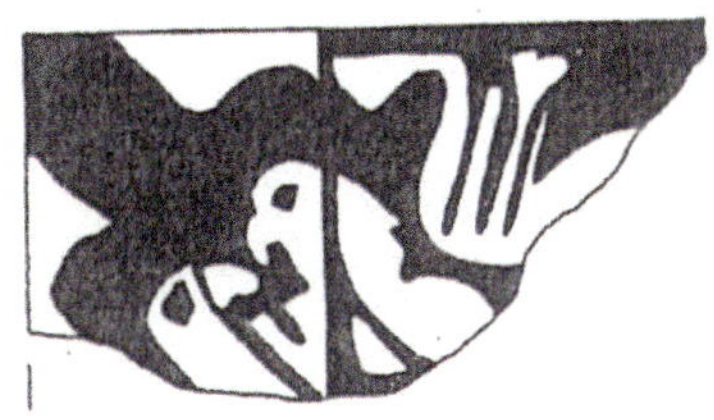 SAME STAMP?
8 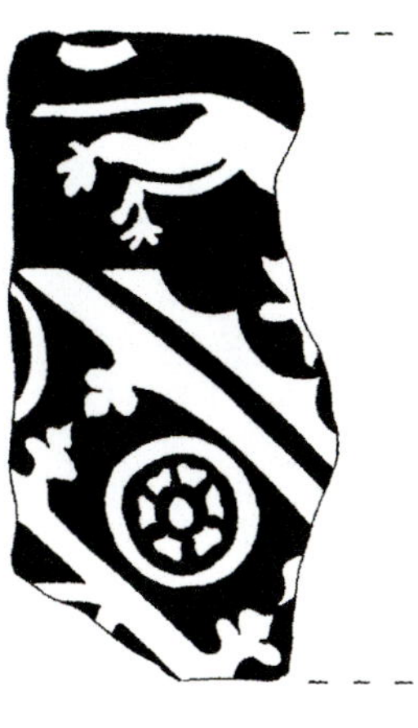	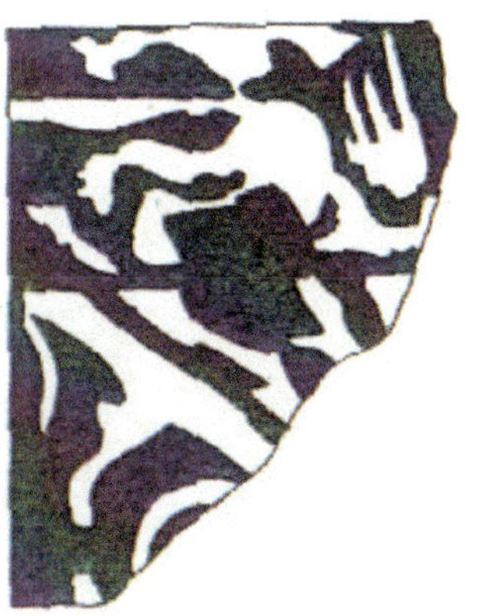 SAME STAMP?

DILWYN

BLACKFRIARS, GLOUCESTER

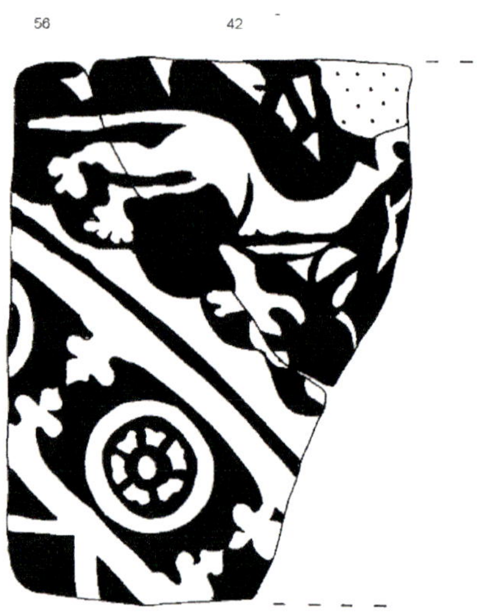

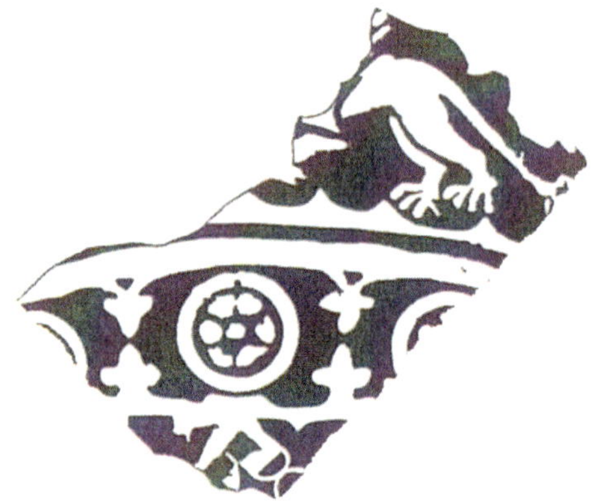

After Eames BM 2938.

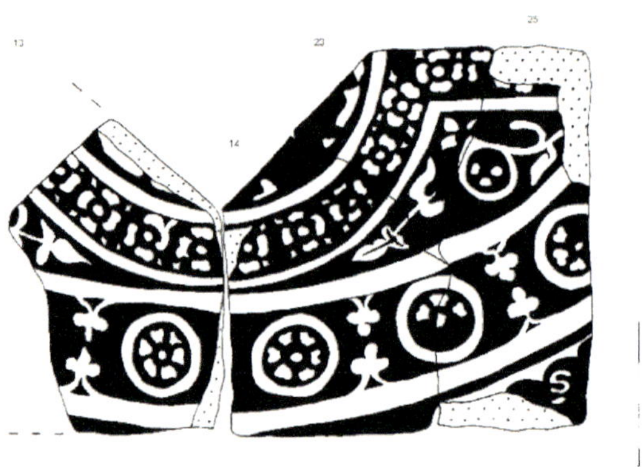

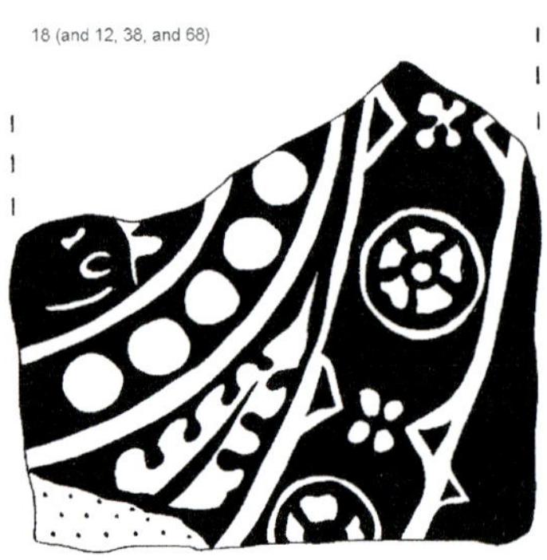

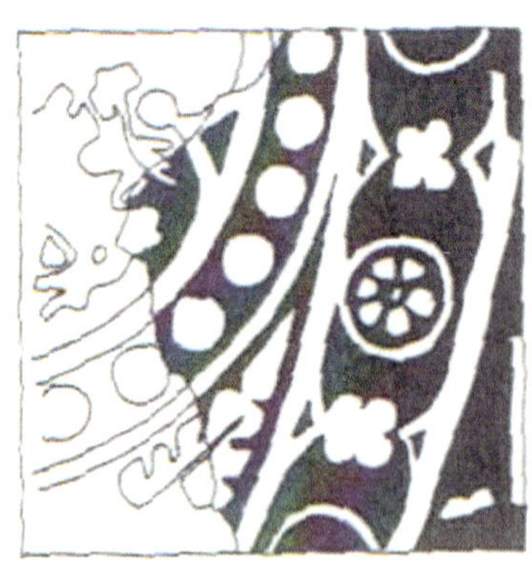

DILWYN

24

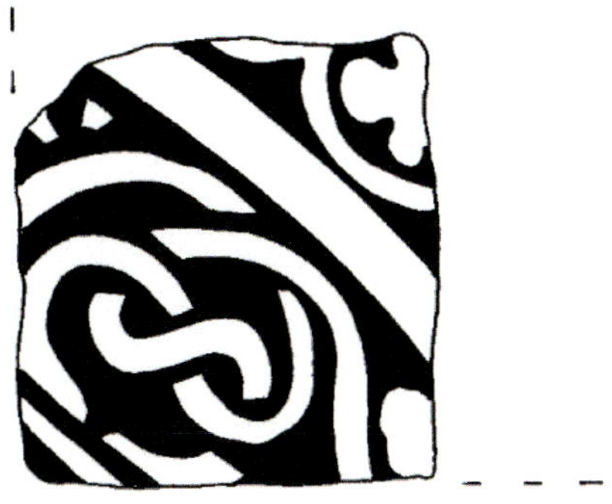

51 52

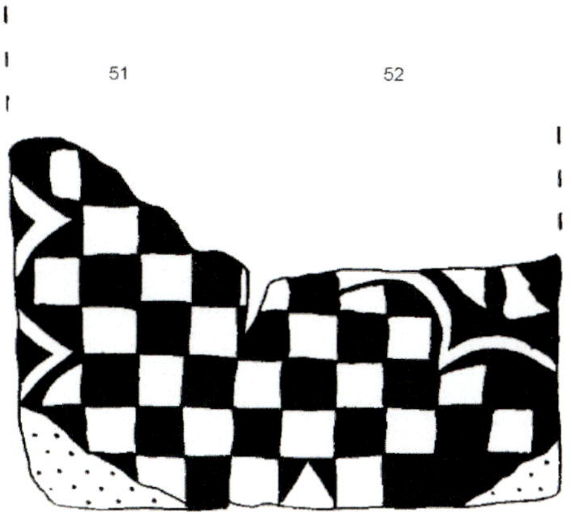

49

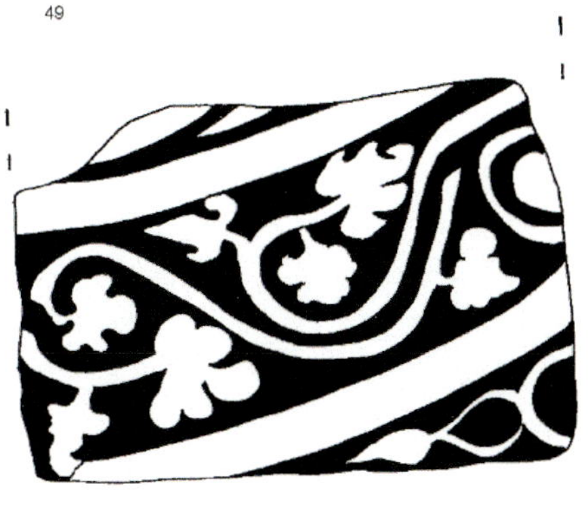

BLACKFRIARS, GLOUCESTER

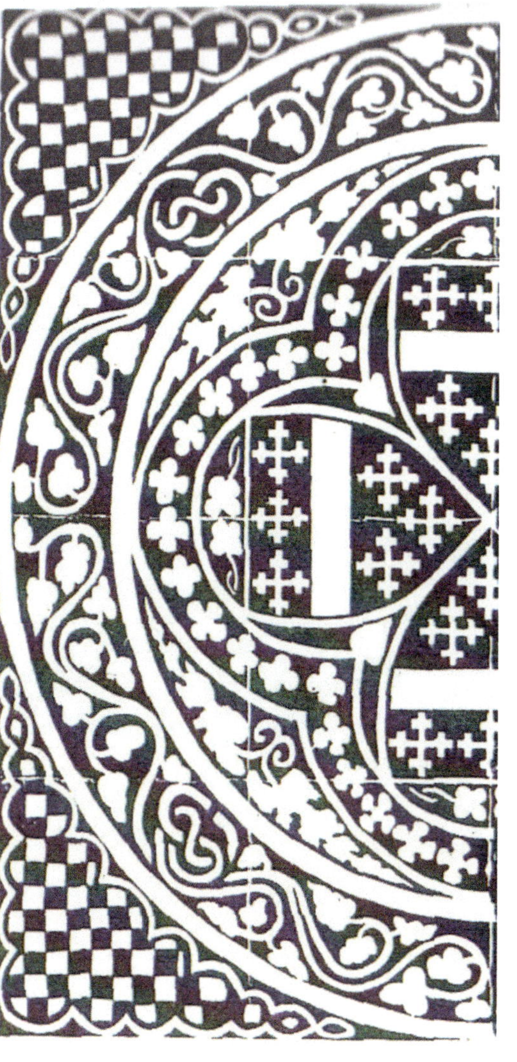

Also listed by British Museum (3053) at Meaux Abbey Mid 13th century

DILWYN

BLACKFRIARS, GLOUCESTER

22

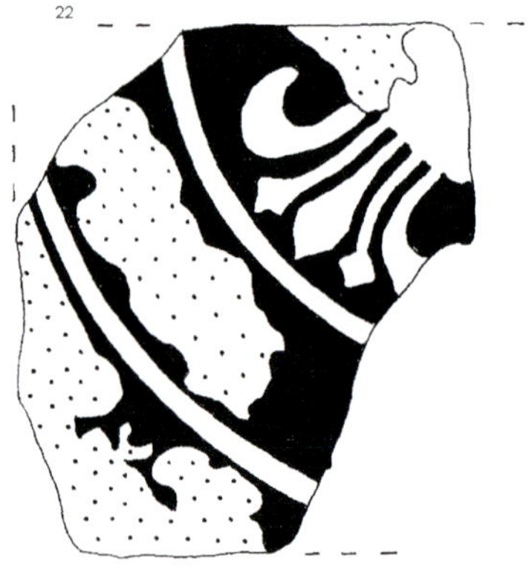

66 (and 63)

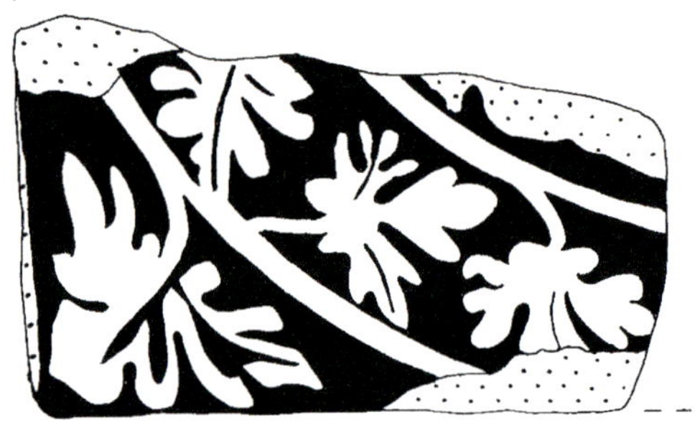

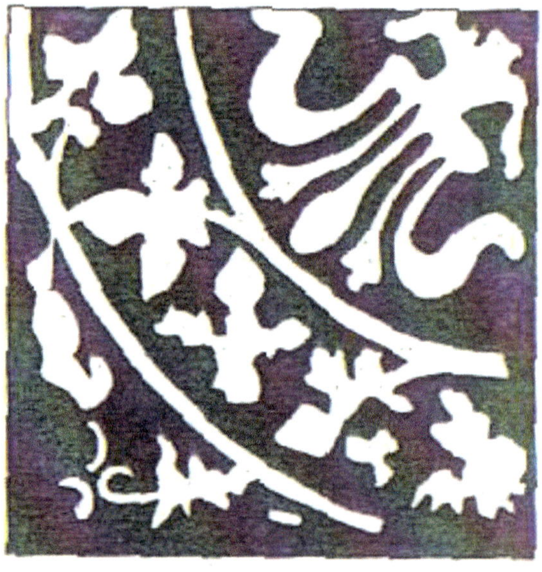

50

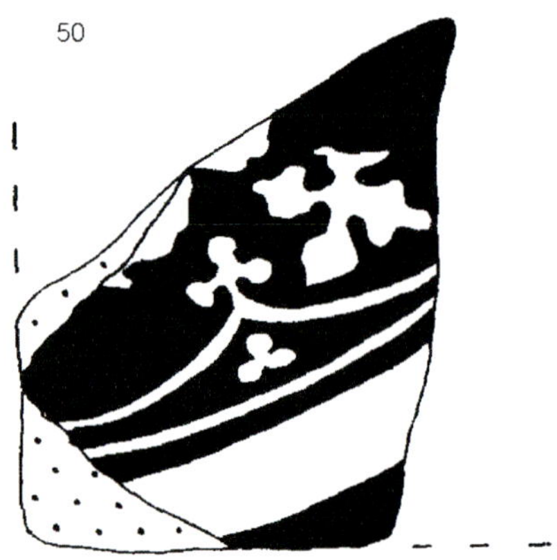

The Dilwyn example lacks the two dots either side of the trefoil compared to this Blackfriars tile. (Also JM Lewis 297a from Carmarthen).

DILWYN | BLACKFRIARS, GLOUCESTER

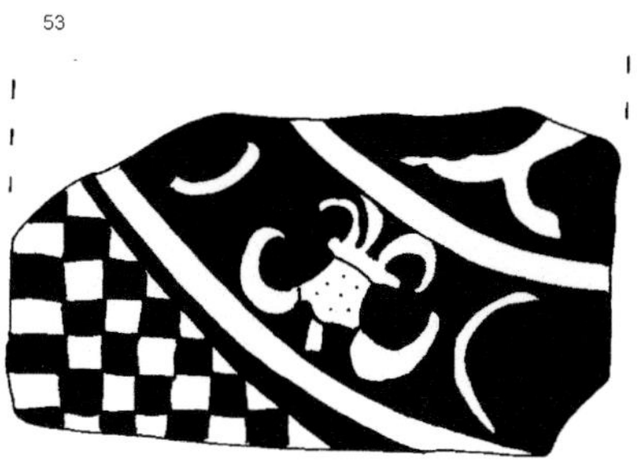

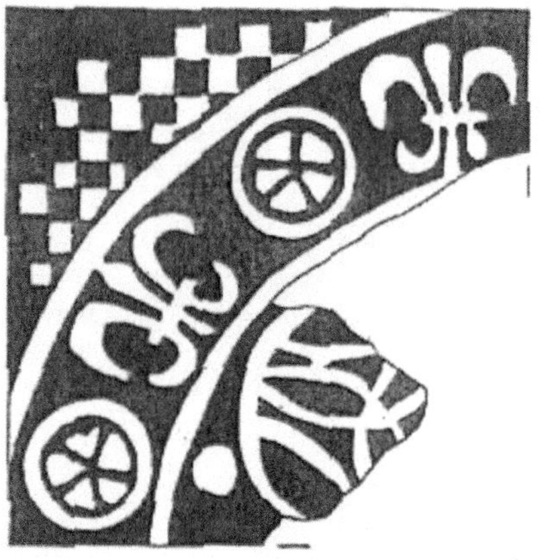

(Also JM Lewis 296).

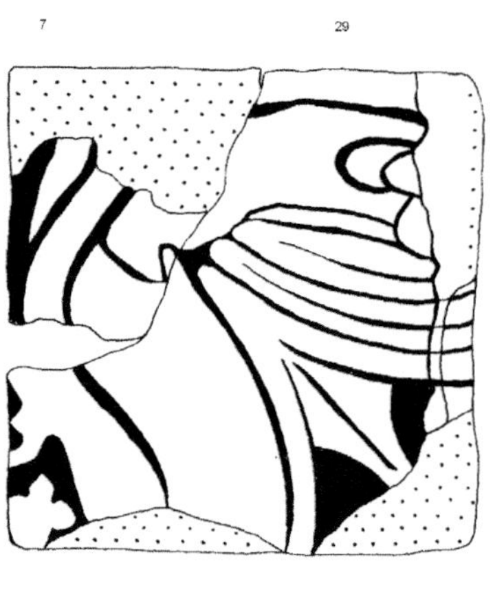

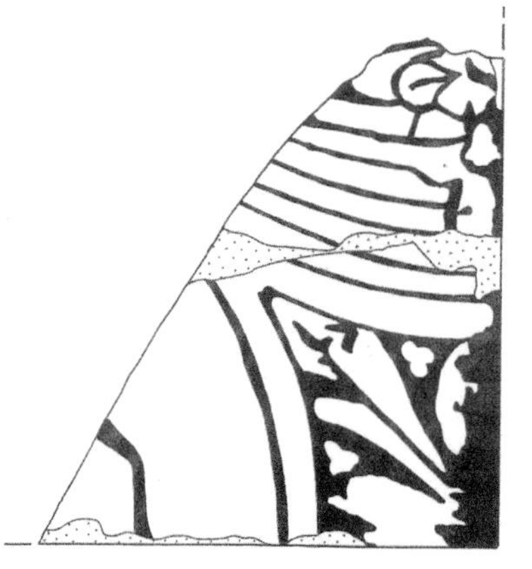

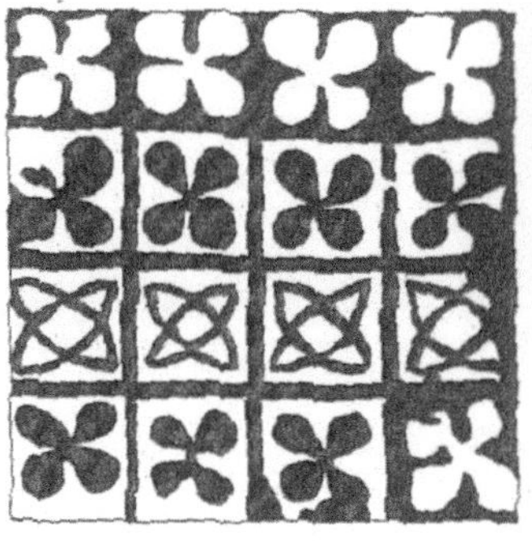

Also at Hall Court, Much Marcle,
Blackfriars Gloucester, Tewkesbury, Bredon.

<table>
<tr><th>DILWYN</th><th>BLACKFRIARS, GLOUCESTER</th></tr>
<tr><td>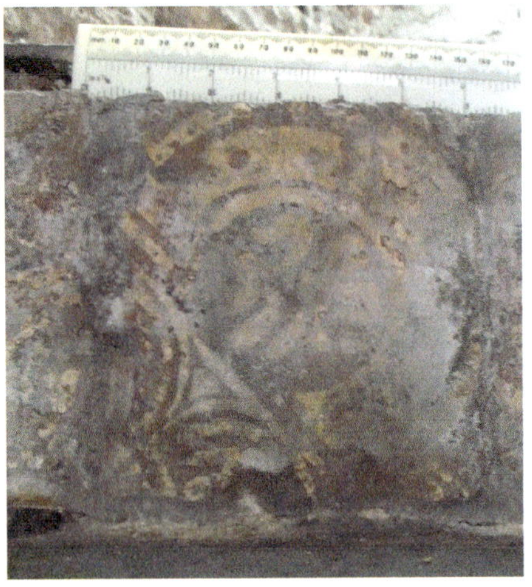</td><td>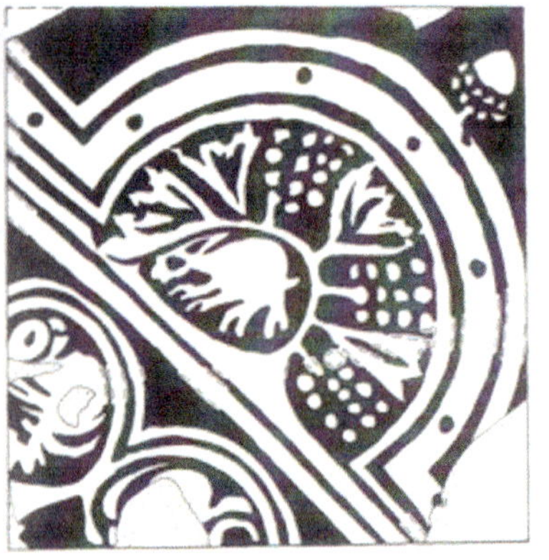
Similar at Blackfriars, Gloucester, but with different centre to roundel.</td></tr>
<tr><td>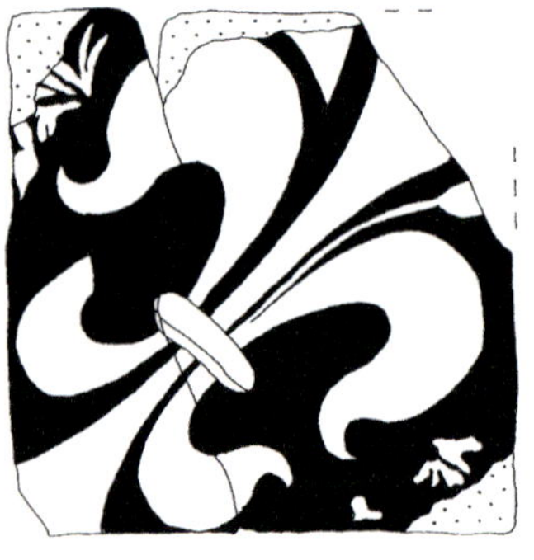</td><td>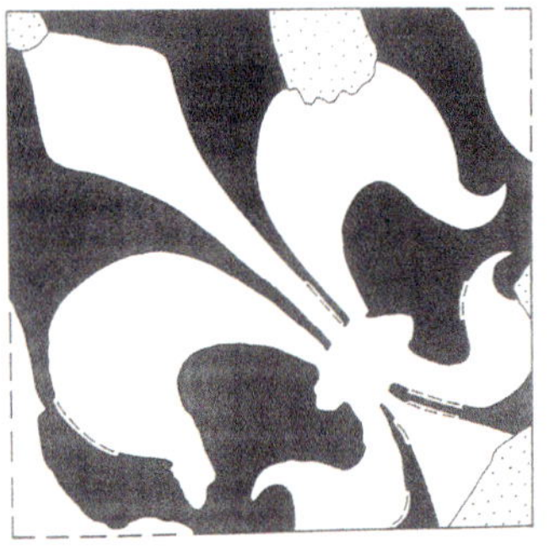
Similar at Blackfriars, Gloucester (not identical) no stamens.</td></tr>
<tr><td>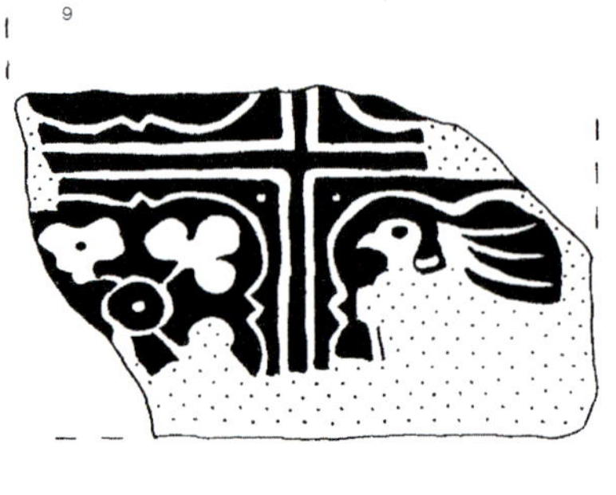
This design has not been identified elsewhere in Herefordshire, but is similar to the example opposite, from Yorkshire.</td><td>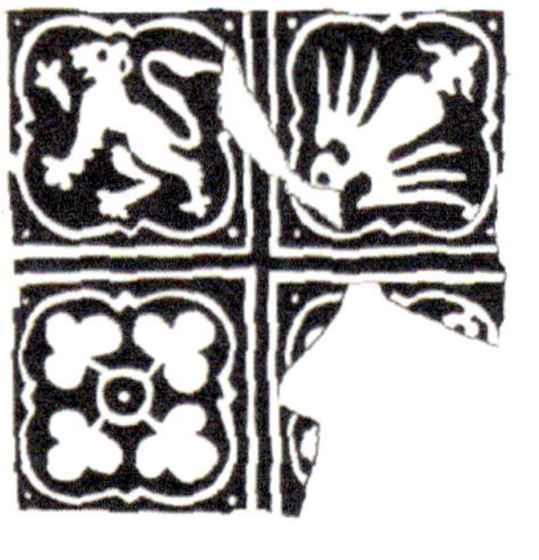
After Eames BM 1886
Byland Abbey, Yorkshire, mid 13th century.</td></tr>
</table>

There were several floortiles depicting armorial shields:

16 and 17 (also 10, 33, 35, 40, 46)

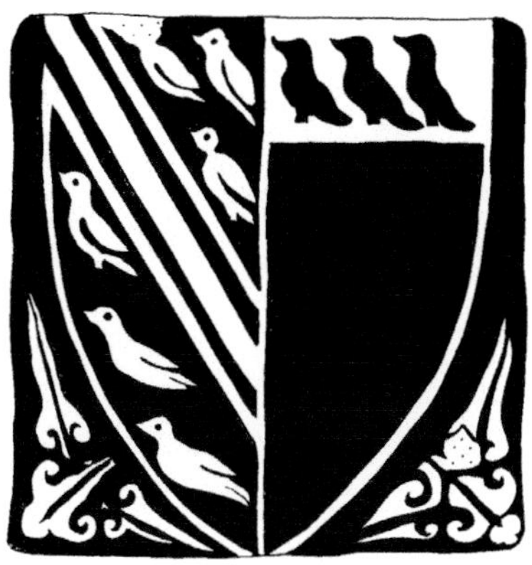

A member of the Chepstow Archaeological Society researched this shield and suggested that this represents the arms of Delabere impaling those of Chabnor.

'per pale: dexter: DELABERE; sinister Chabnor Blazon from Burke's General Armory.

DELABERE: Azure a bend Argent cotised Or between six martlets of the last.

CHABNOR: Gules on a chief Argent three martlets sable.

(There was a marriage between the two families during the 13th or 14th centuries.)

3 6

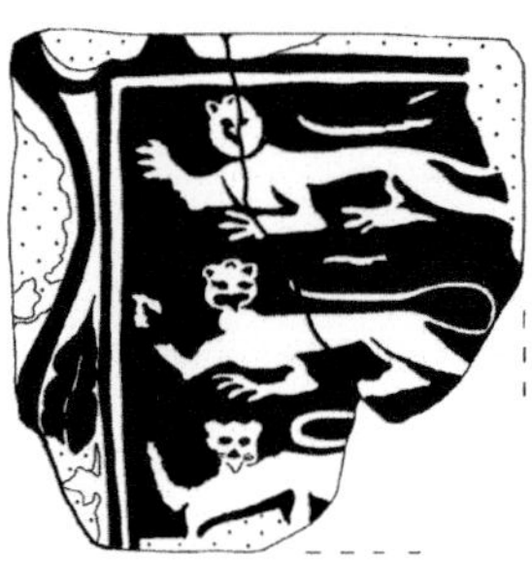

Possibly:
A variant of the 3 lions Royal coat of arms.

Gules, three lions passant guardant or, a label azure.

Used by the eldest sons of the king during their father's lifetime (ie Edward I, II, or III).

2

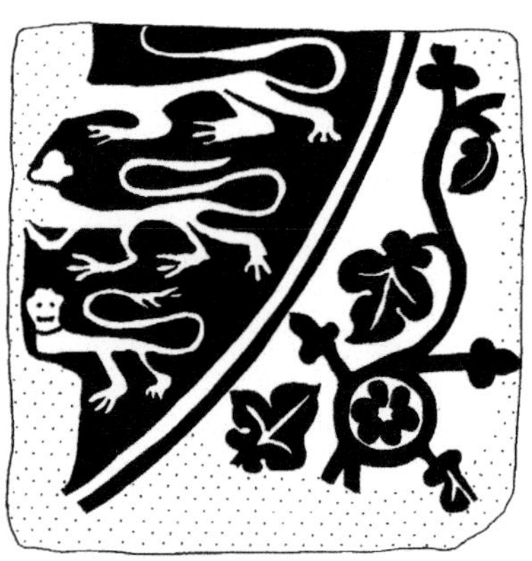

Unknown
Likely to be the three lions of England.

30

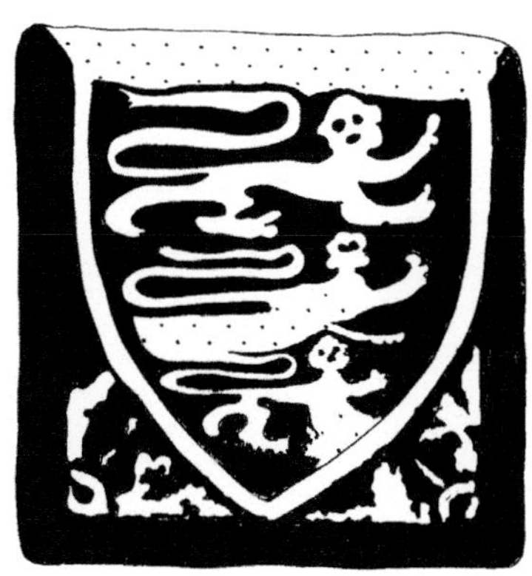

Unknown
Three lions but facing sinster?

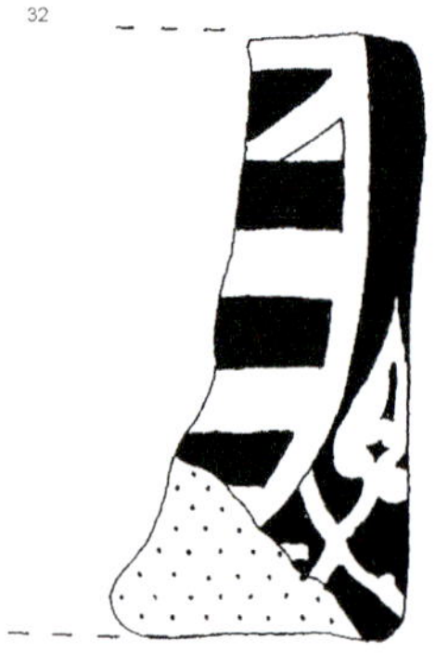

Probably a Mortimer family?

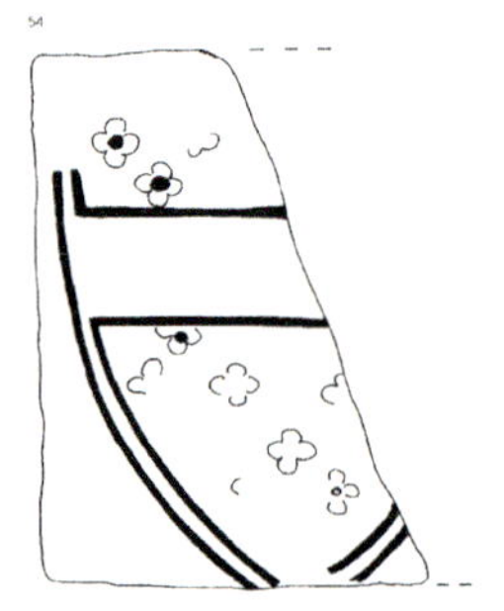

Unknown.

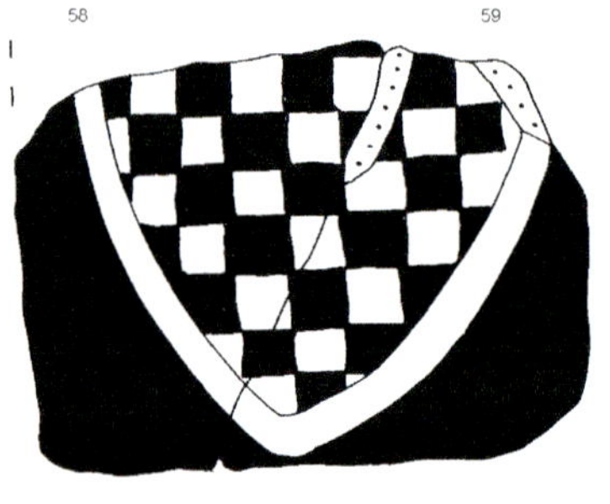

Possibly Clifford?

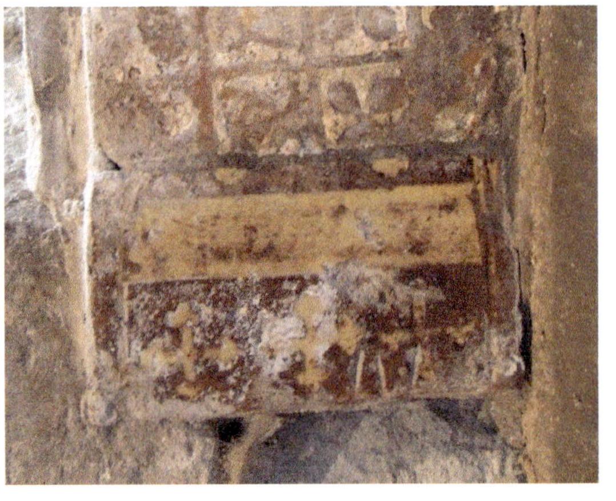

Top part of heraldic shield – white band with 3 cross crosslets? below?
Possibly Beauchamp?

Photographs from visit to the church:

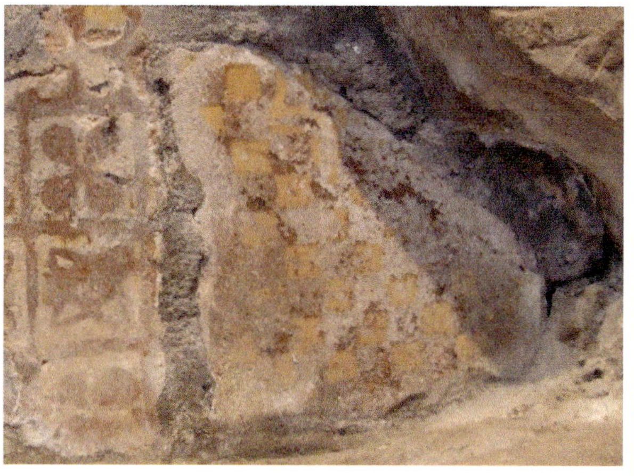

Very small chequers, cf Leominster etc.

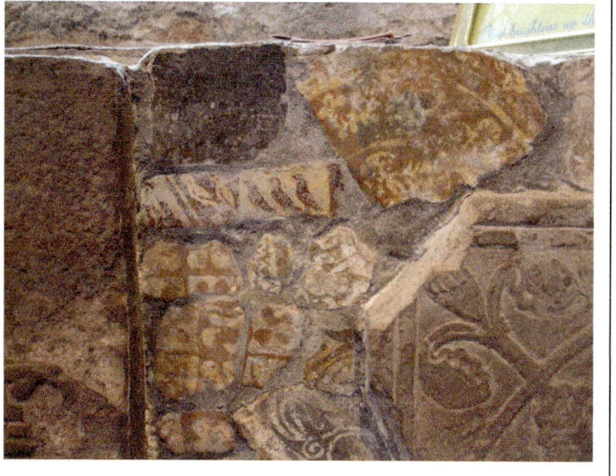

What is this top right?

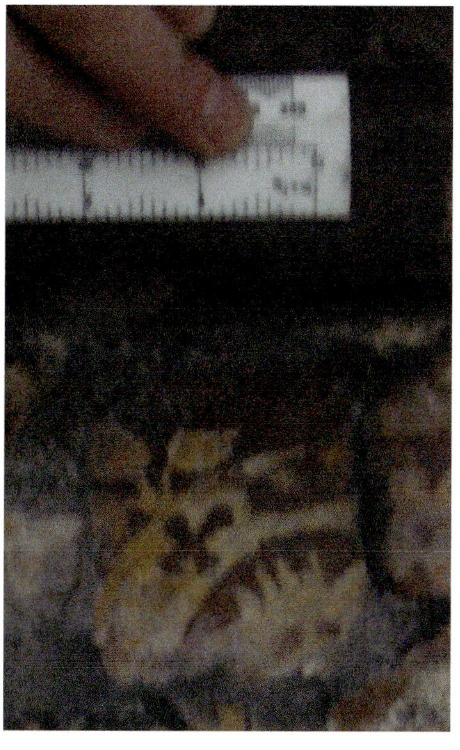

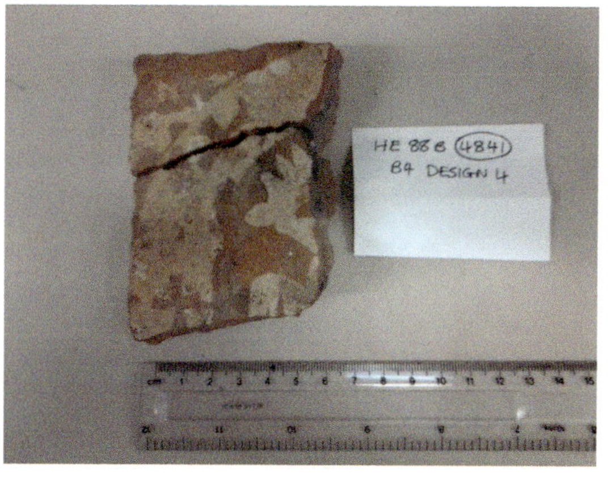

Fragment from Deen's Court, Hereford
(in Hereford Museum collections).

Unidentified:

65 (and 19 and 28)

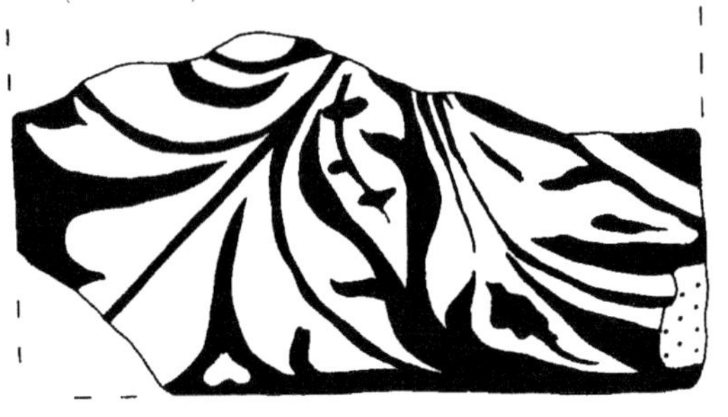

48

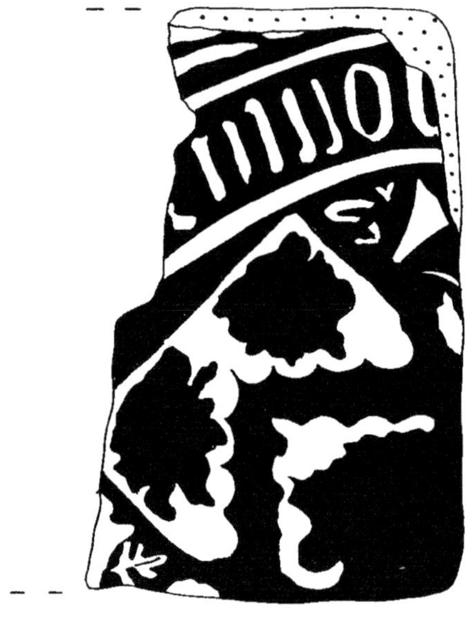

43 (and 64 and 37)

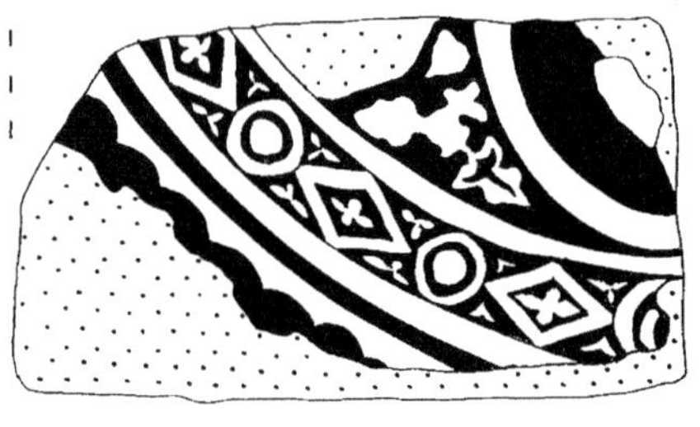

Goodrich Castle

Goodrich Castle is now in the hands of English Heritage. I understand there was some archaeological work at the Castle in the 1990s which produced 47 fragments of floor tile, 34 of which were decorated with nine identifiable designs. Alan Vince and Ron Shoesmith reported on the tiles in great detail in the AVAC Report of 1998; their prime interest is with the origins of the ceramic material whereas I can only deal with the designs. I have not seen these tiles and have not been able to establish where they are curated.

The designs were not drawn but I list below the text descriptions given. The drawings are what I have deduced the tiles would have looked like from these written descriptions.

'Border tile with 5 or 6 circles between two parallel lines.'	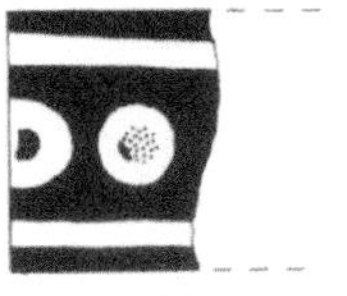 90	This example is J M Lewis 90 Group 11 (SE Wales and Bristol area). Moynes Court, Chepstow.
'Crossed vessicas with circles in the spandrels and a four-lobed flower in the centre.'	 14	This example from Worcester Cathedral has 8 lobes in the centre. The Parker-Hore archive (Tileweb) shows a similar design attributed to Goodrich (see below).
'Four fleurs-de-lis surrounding a square containing four squares.'	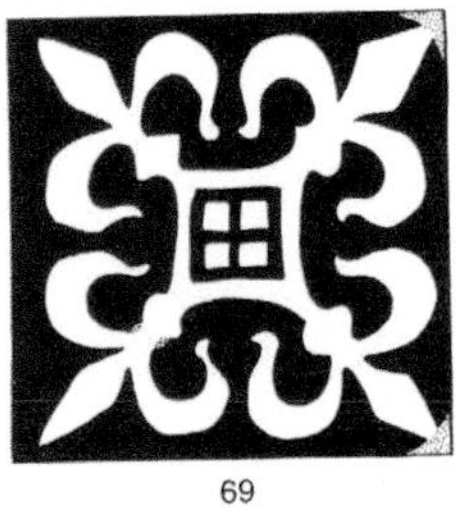 69	JM Lewis' Group 11 (SE Wales and Bristol area) BM 2195 This one from Tintern Abbey.
'Two birds within a circle with their backs towards a tree. Quarter flowers in the corners.'	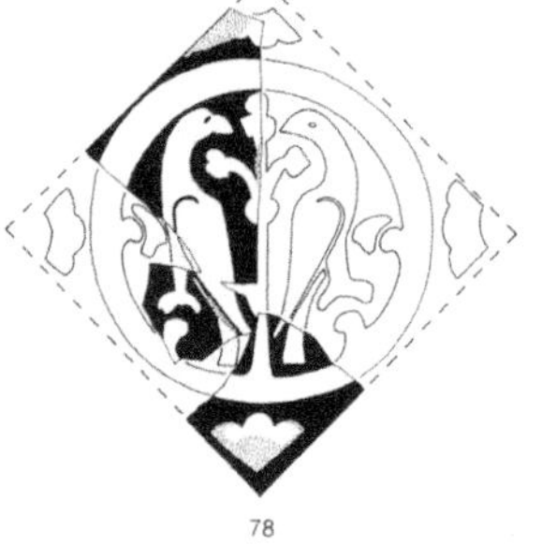 78	JM Lewis' Group 11 (SE Wales and Bristol area) BM 1984 This one from Moyne's Court and Tintern Abbey.

'A griffin within a circle with fleurs-de-lis in the corners.'	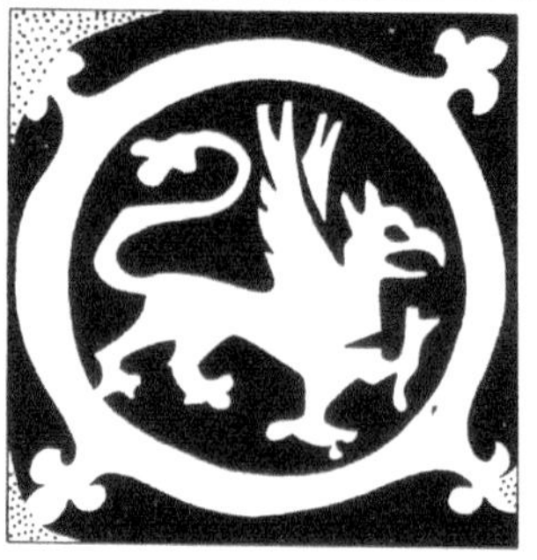 33	JM Lewis' Group 8 (Wessex type) This one from Monmouth.
'A dragon facing left within a circle with fleurs-de-lis in the corners.'	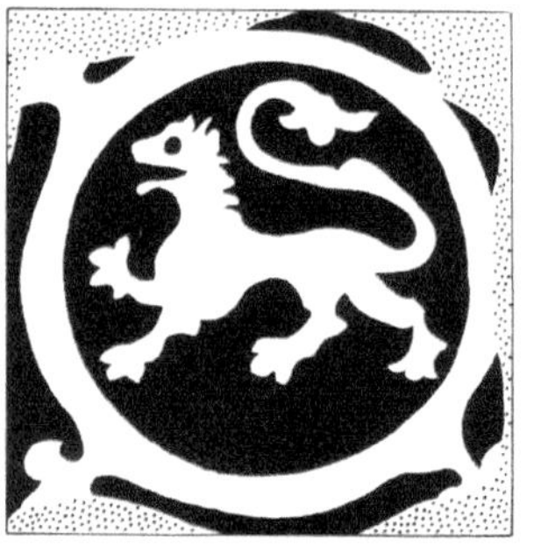 34	JM Lewis' Group 8 (Wessex type) This one from Monmouth. This is a lion (not a dragon) but JM Lewis mentions a possible griffin found at Monmouth. Similar tiles are found at Netherwood, Thornbury; Wigmore Abbey; Worcester Cathedral; and the Lady Chapel of Hereford Cathedral.
'A lion facing right within a circle with stylised fleurs-de-lis or flowers in the corners.'	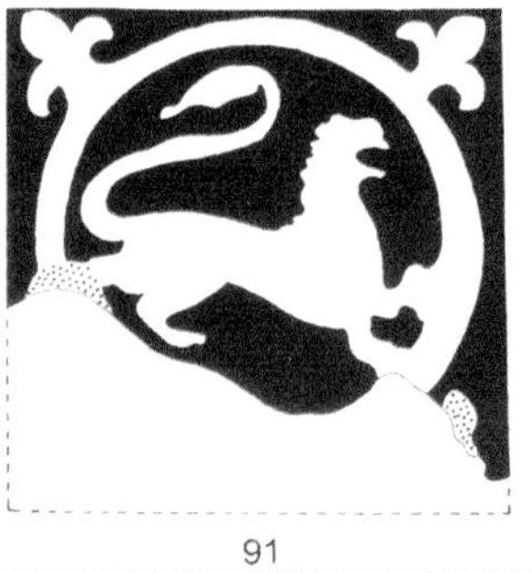 91	JM Lewis' Group 12 (Severn/Bristol Channel area) This one from Kidwelly Castle.
'Four tile pattern. The complete pattern consists of a dotted band quatrefoil enclosing four fleurs-de-lis with a dotted circular band at the centre and flowers in the corners.'		
'Unidentified fragment probably of a mythical beast within a circular band, quite possibly one of those already from Goodrich.'		

There are three tiles recorded from Goodrich Castle in the Parker-Hore archive collection on the Tileweb website:

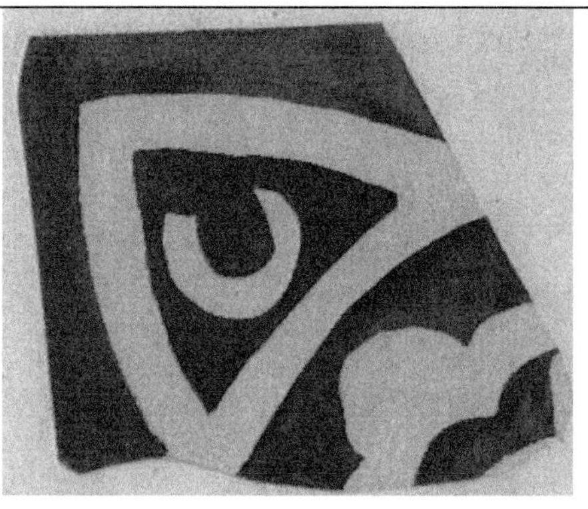	Likely to be the crossed vessicas referred to above.  14 Worcester Cathedral.
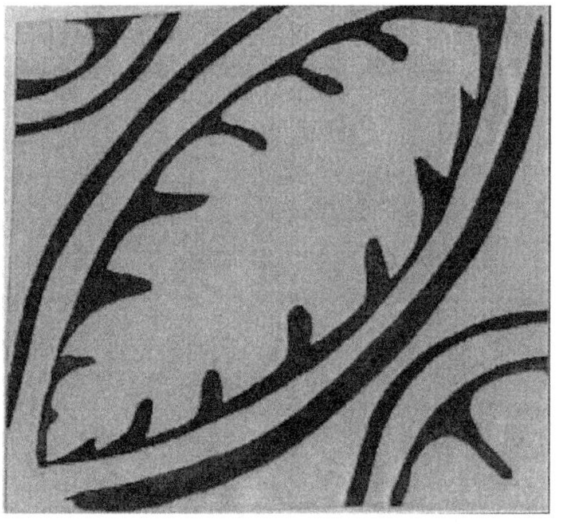 Unknown.	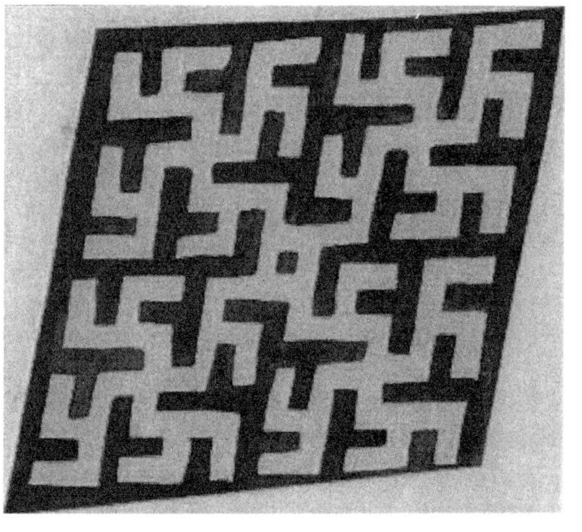 Unknown.

Hereford – Church of All Saints

(Photographs courtesy of Laurence Keen)

Set in the wall of the eastern pier of the north aisle is a panel of ten tiles.

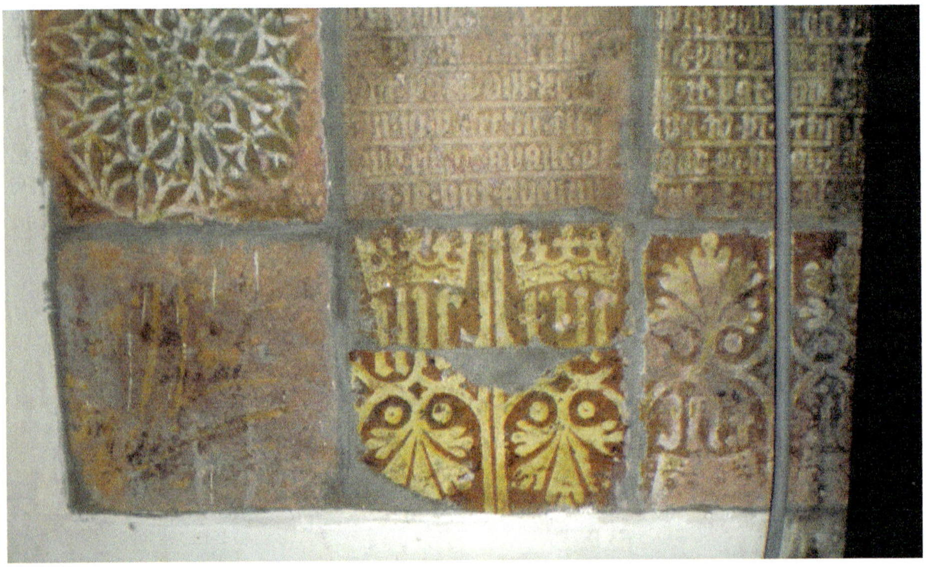

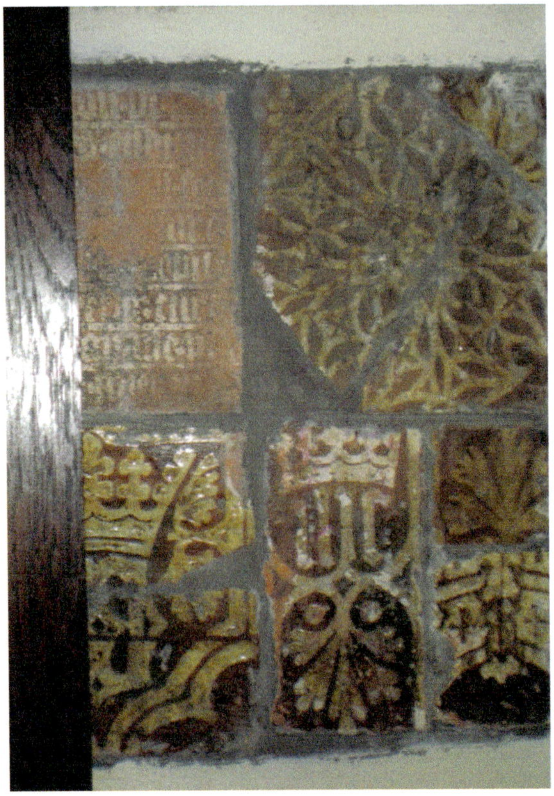

Three feature a black letter inscription (JM Lewis 325); two feature the *ihc/xpc* crowned letters; one depicts a crowned M; two are an as yet unidentified sunburst design, and one is so badly worn it is impossible to identify a design. The tile in the bottom right hand corner is in three pieces and comprises fragments of both 331 and 332.

325 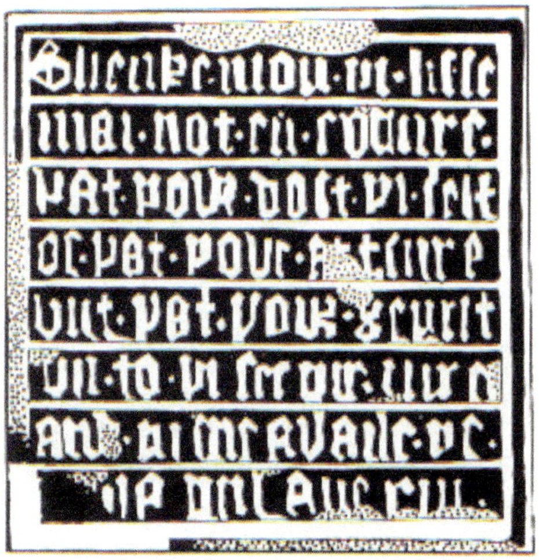 325	JM Lewis Group 21 (Great Malvern c. 1453) (Probably made at the Cadogan House kiln in Monmouth). Slip-on-impression. Black-letter inscription in eight lines, probably reading: *Thenke mon thi life/ mai not eu endure/ That thou dost thi self/ of that thow art sure/ availe the/ hit is but aventure.* This has been interpreted as a warning to executors, for which there are 15th century parallels. Also at: Monmouth, Raglan.

311 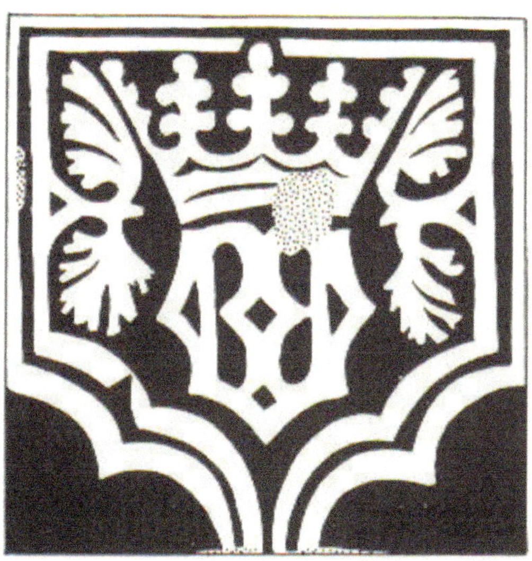 311B	JM Lewis Group 21 (Great Malvern c. 1453) (Probably made at the Cadogan House kiln in Monmouth). Slip-on-impression: crowned M within double border, for Henry VI's queen, Margaret of Anjou (d.1482). BM Design 1421 Also at: St Michael's Croft, Stretton Sugwas, Raglan, Gloucester Cathedral, Great Malvern Priory.

331

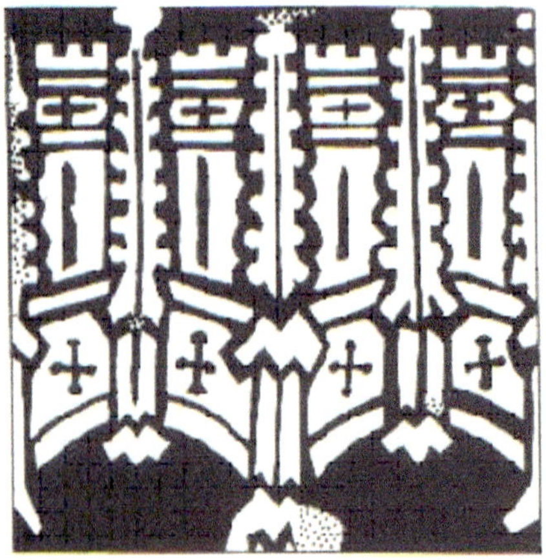

331

JM Lewis Group 21 (Great Malvern c. 1453) (Probably made at the Cadogan House kiln in Monmouth).

331-4 Square wall-tile series based on tabernacle-work, made originally for Great Malvern Priory.

331 Slip-on-impression. Turrets and pinnacles.

BM Design 2577 (Great Malvern Priory).

Also at: Stretton Sugwas, Monmouth, Raglan.

332

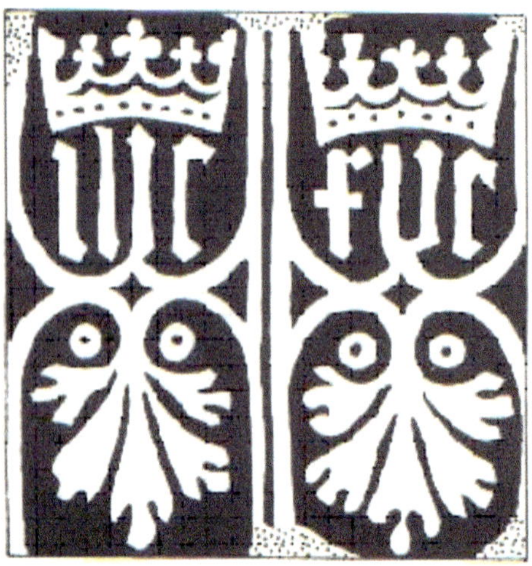

332

JM Lewis Group 21 (Great Malvern c. 1453) (Possibly made at the Cadogan House kiln in Monmouth).

332 Slip-on-impression. Two-light opening with pendant leaf motifs, and with crowned monograms above: *ihc* and *xpc* (Jesus/Christ).

BM Design 1420 (Great Malvern Priory, Lenton Priory kiln site).

Also at: St Michael's, Croft, Stretton Sugwas; Abergavenny; Llangattock-nigh-Usk, Monmouth.

Hereford Cathedral – Booth Chapel

These tiles are in an area not usually open to the public, in the Booth Chapel over the North Door. They have been relaid in a random fashion and show varying degrees of wear.

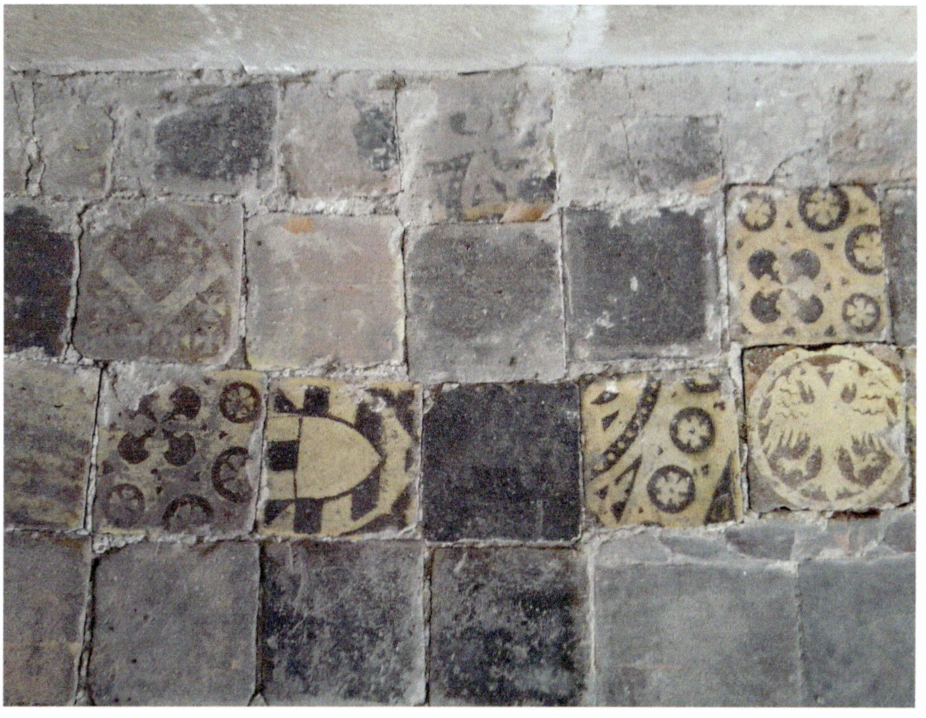

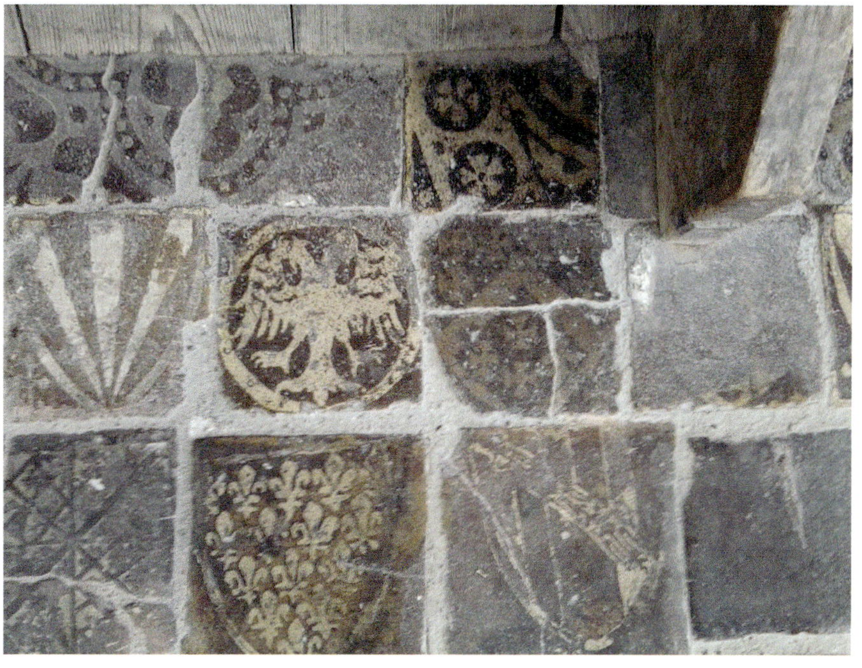

It is difficult to use the photographs of this site to show individual tiles, so I will use drawings of the designs found in the Booth Chapel where these are available, but I should make clear that these are not drawings of the actual tiles in the Booth Chapel.

Three designs (271-4) (also found at Chepstow Priory) show a larger proportion of the surface area with the cream infill glaze than is usual, and would have had a marked visual impact.

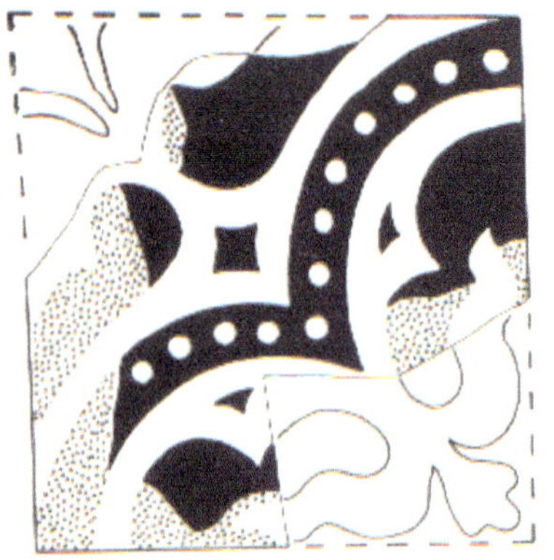

271

JM Lewis Group 20 (Droitwich).

271
Stamp on slip. 4 tile design: quatrefoil outlined by reserved band with pellets, with fleur-de-lys motif springing from cusps; outside, springing from each angle, a further fleur-de-lys motif.
Compton 180 (Worcester Cathedral transept: cf BM Design 2647).

Chepstow Priory.

272

273

JM Lewis Group 20 (Droitwich).

272-3
16 tile design: circular band decorated with reserved roundels containing five-petalled rosettes, surrounding a quatrefoil outlined by a reserved band with pellets, and with internal tracery and foliage.

Chepstow Priory.

There is a similar floortile in the collections of Hereford Museum found from the Vicars Choral, Hereford Cathedral.

274

JM Lewis Group 20 (Droitwich).

274
4 tile centre of 16 tile design; including reserved roundels with five-petalled rosettes.
Cf 272 Compton 180 (Worcester Cathedral transept).

Chepstow Priory.

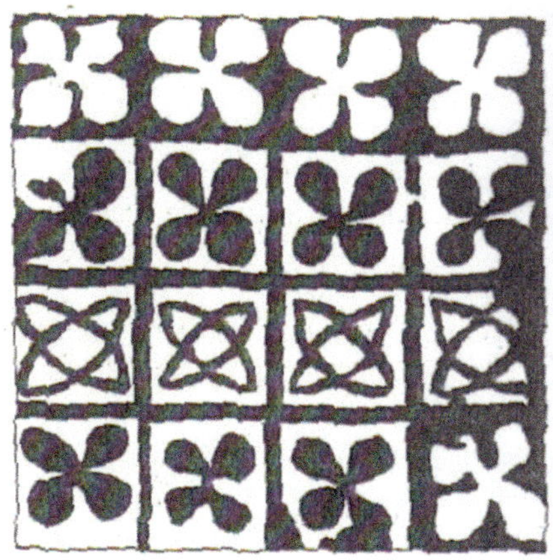

This drawing is from Blackfriars, Gloucester.

Also at Dilwyn and Much Marcle Hall Court.

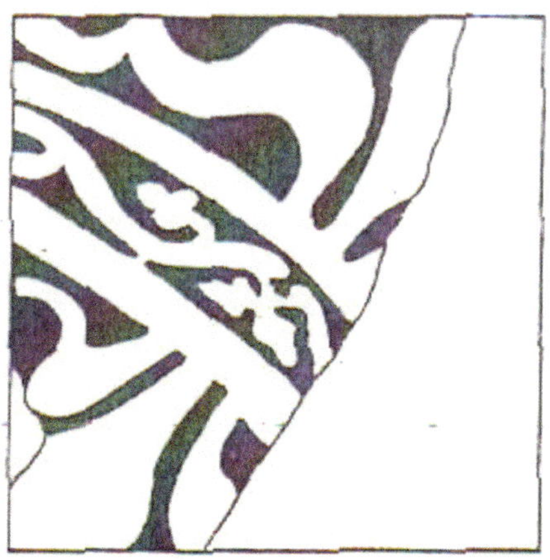

This drawing from Blackfriars, Gloucester, forms a 9-tile design.

Also at Lower Breinton.

This drawing is from Blackfriars, Gloucester.

Also at Much Marcle, Hall Court.

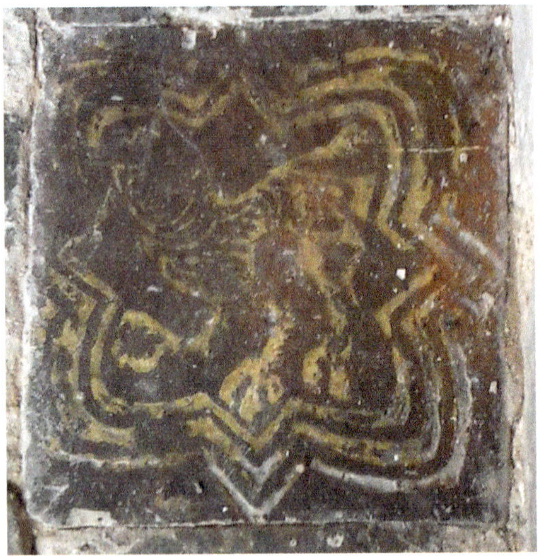

Blackfriars, Gloucester

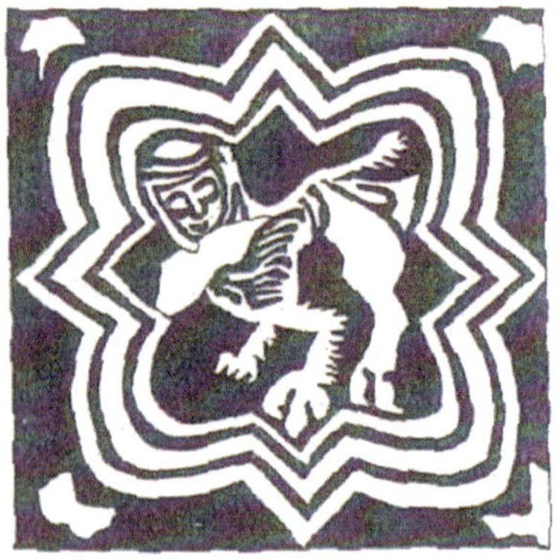

(different head and front right foot).

Similar tiles have a cat/lion face or a nun; the Hereford example appears to be a cat.

Also at Castle Frome, and Limebrook Nunnery

After Eames BM 2713 Ulverscroft Priory (Nottingham) 15 century.

There is a very worn example of this design; also a similar design with a stylized fleur-de-lys within the corner roundel.

A tile from Leominster Old Priory (GS11) shows this same distinctive border, but with a six lobed flower within the roundel.

There is a very worn tile featuring this 'Martin me fecit' text, as found at Abbey Dore:

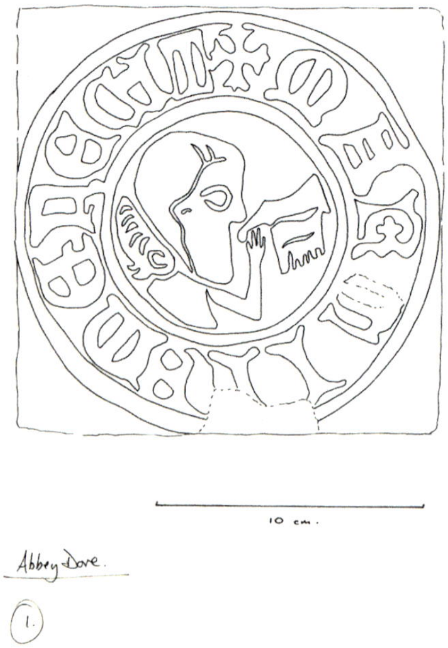

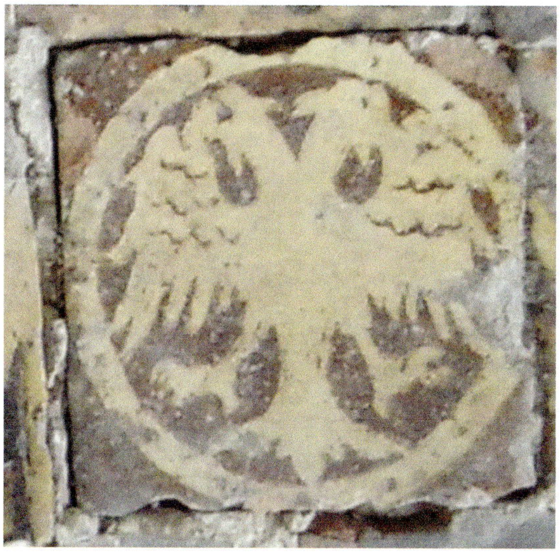

Double-headed eagle within a circle with pellets: Also at Monkland.

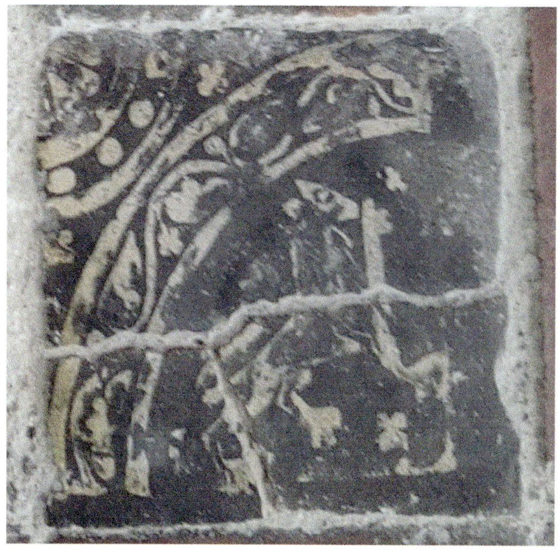

Unknown.

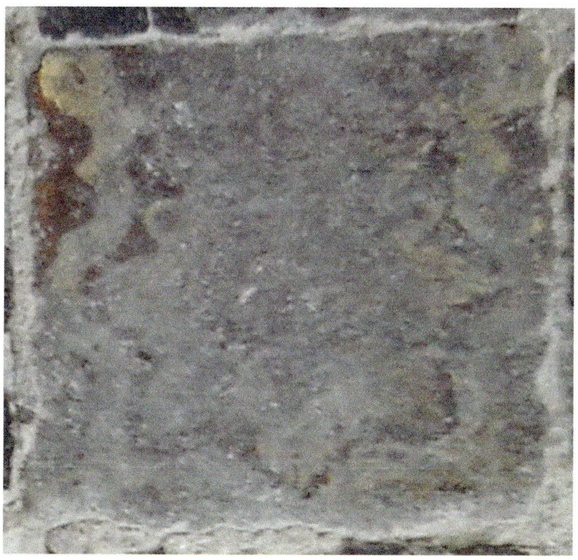

Unknown.

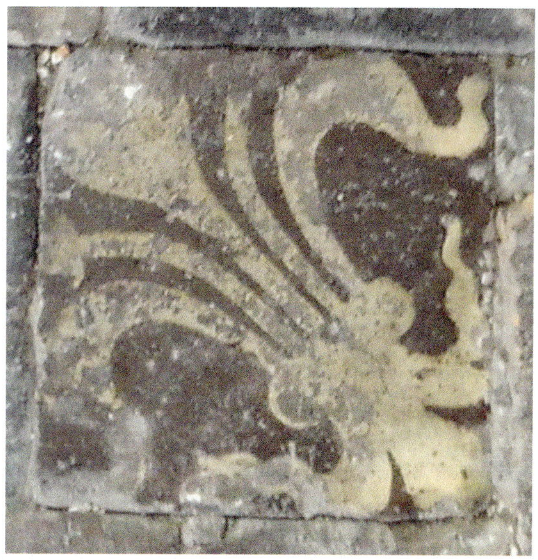

Unknown.

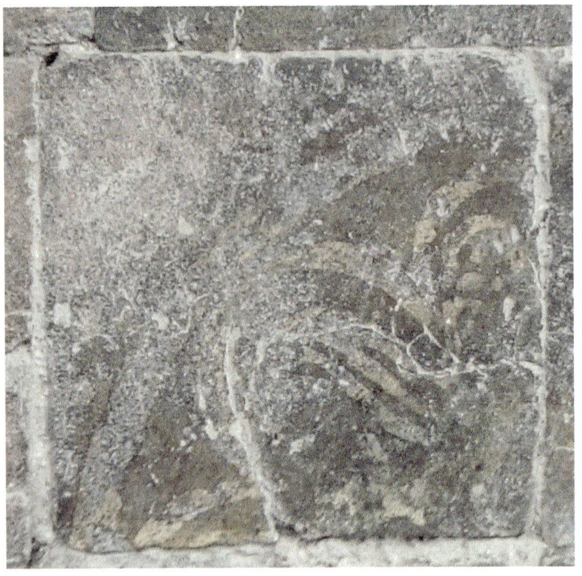 Unknown.	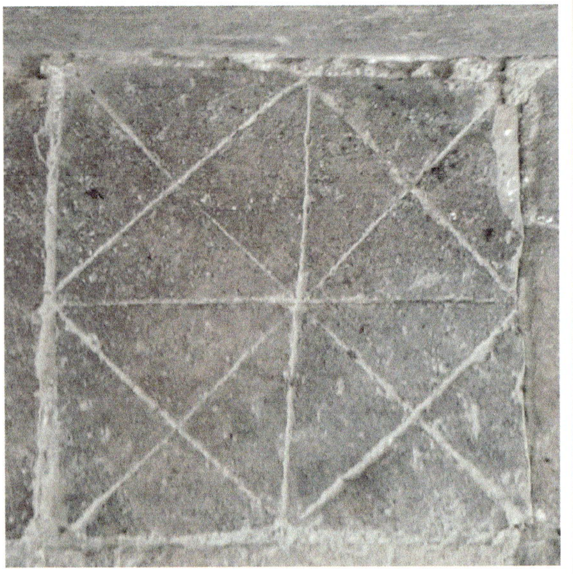 Unknown.

Armorial shields:

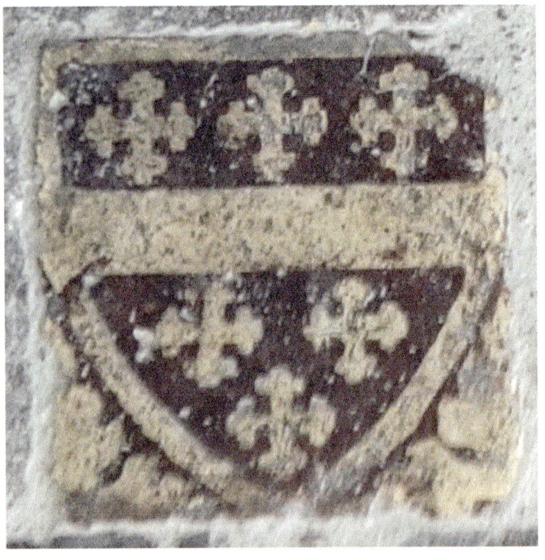 *Gules, a fess between six crosses crosslet or* Beauchamp Earls of Warwick.	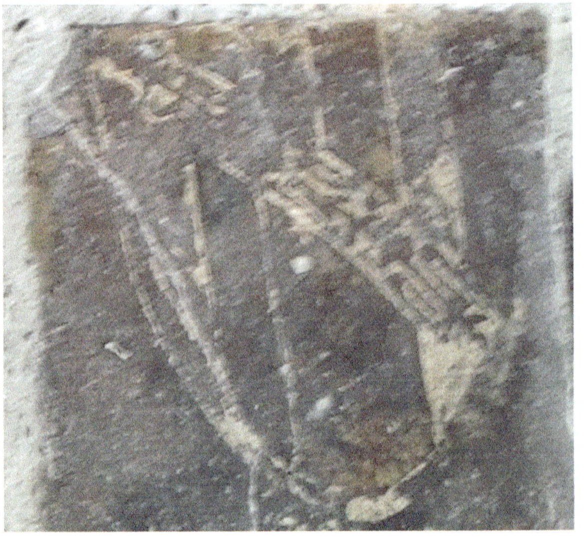 *Paly azure and argent, on a bend gules, three eaglets or, in chief a mullet.* William de Grandison.

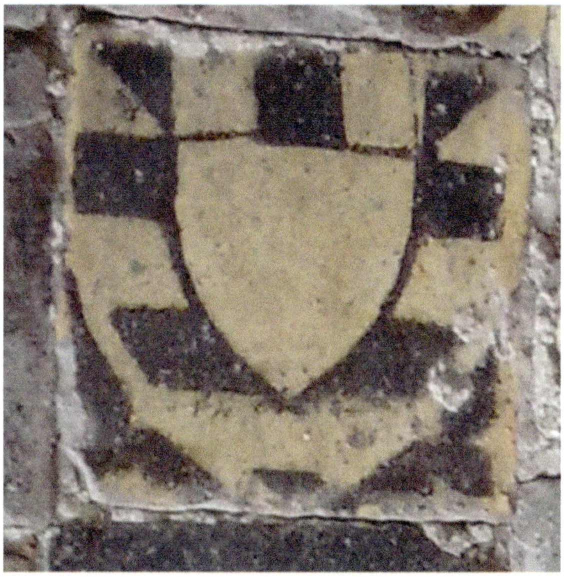

Barry of six azure and or, on a chief azure two palets between two based esquires or, overall an escutcheon argent.
One of a group of the family Mortimer of Wigmore.

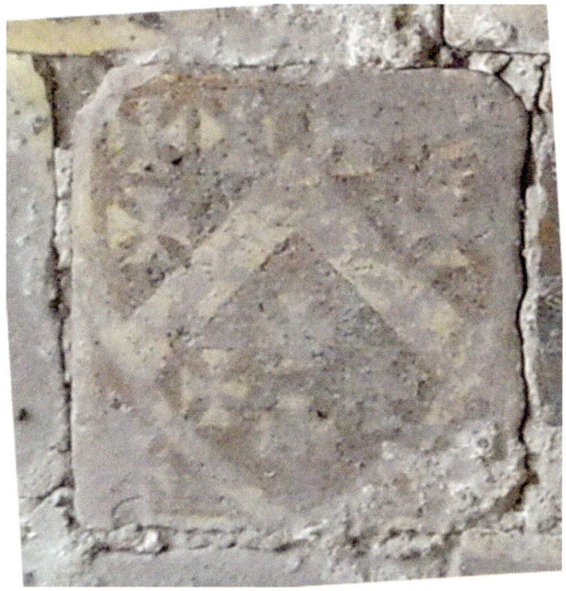

Gules, a chevron between nine crosses pate argent
Thomas Lord Berkeley.

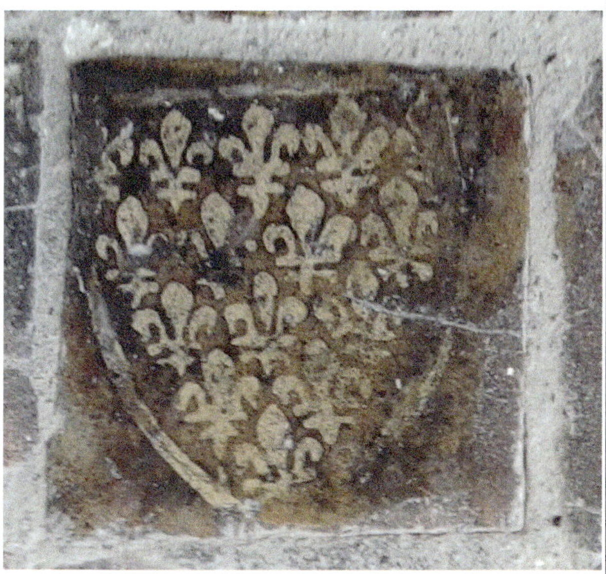

Azure, semee-de-lis or
The arms of France (ancient) adopted by Philip II 1120-80.

Possibly also at Netherwood, Thornbury.

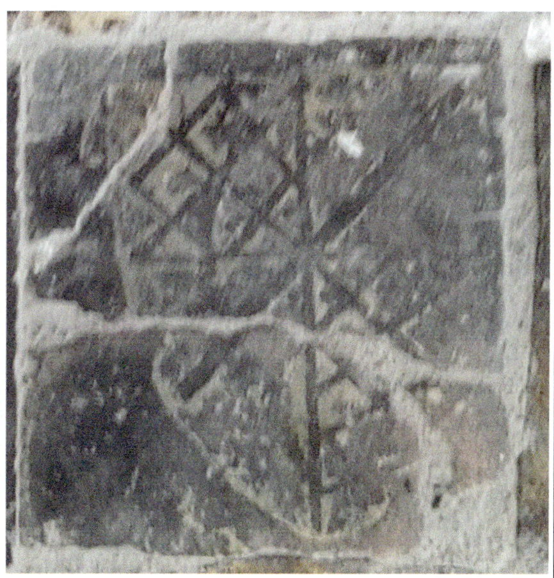

Unknown.

Hereford Cathedral – Lady Chapel

Photographs courtesy of Laurence Keen

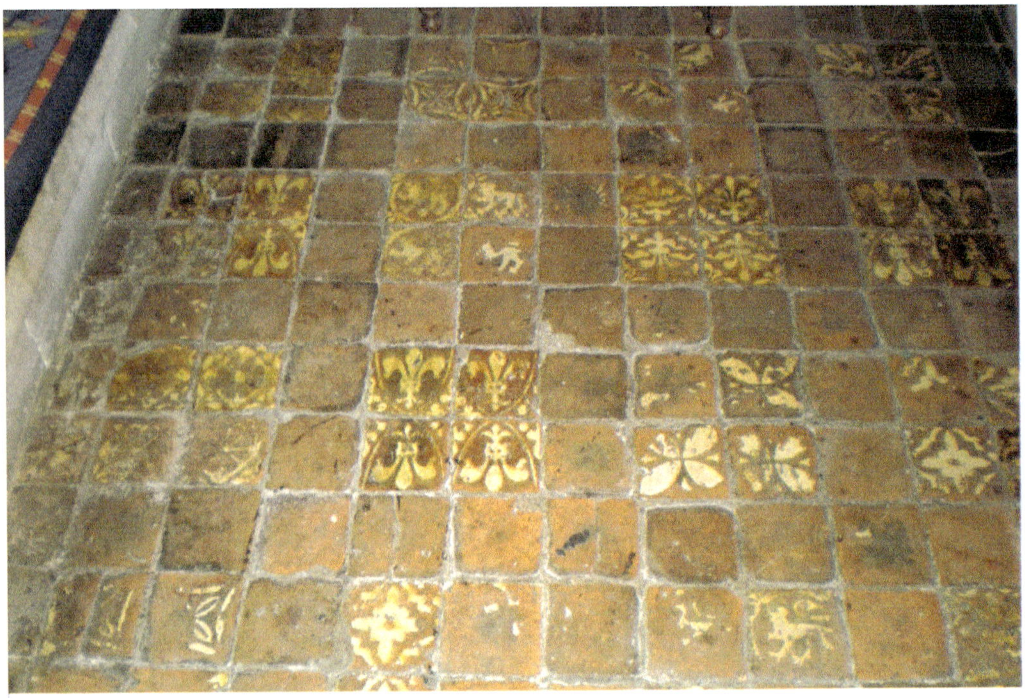

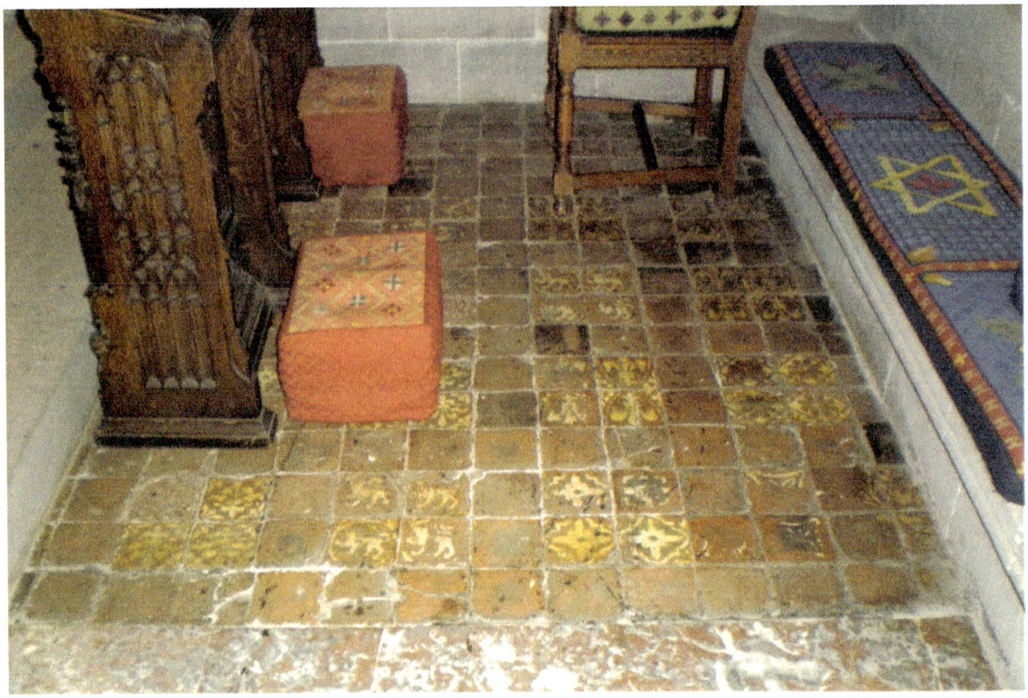

It should be noted that 6 of these 8 designs are also to be found at Wigmore Abbey; 6 at Worcester Cathedral, and 3 at Netherwood, Thornbury.

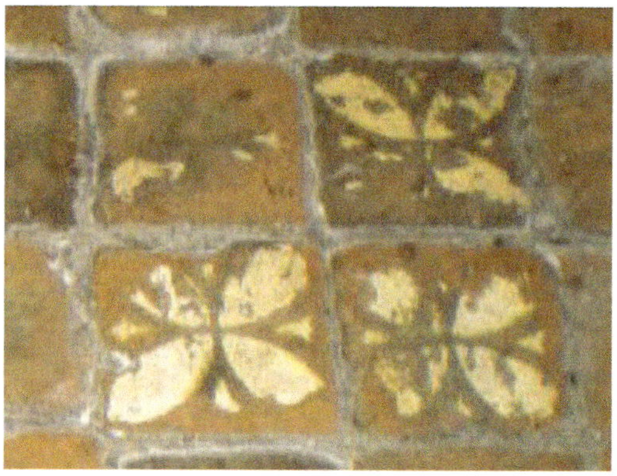 Also at Wigmore Abbey, (similar at Worcester Cathedral).	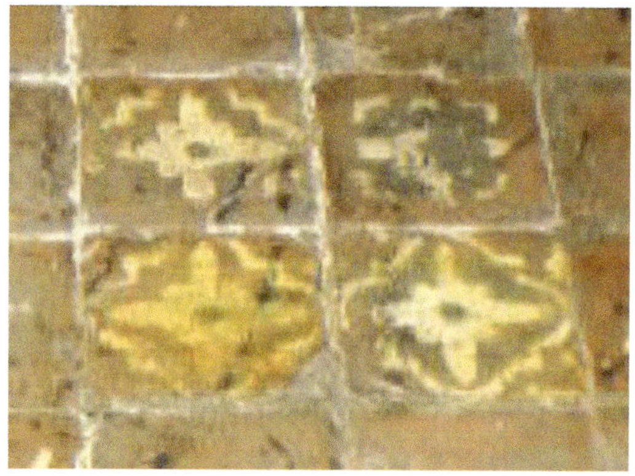 Also at Wigmore Abbey and Worcester Cathedral.
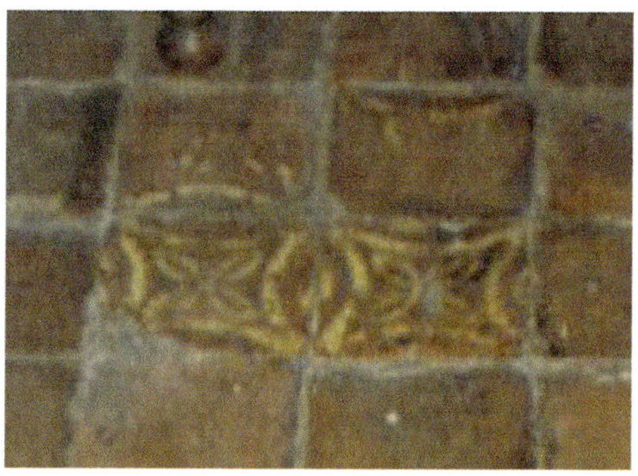 Also at Wigmore Abbey and Worcester Cathedral.	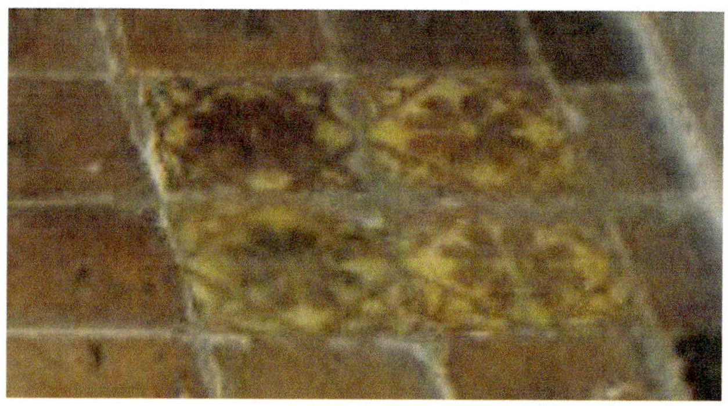 Also at Wigmore Abbey, Netherwood, Thornbury, and Worcester Cathedral.
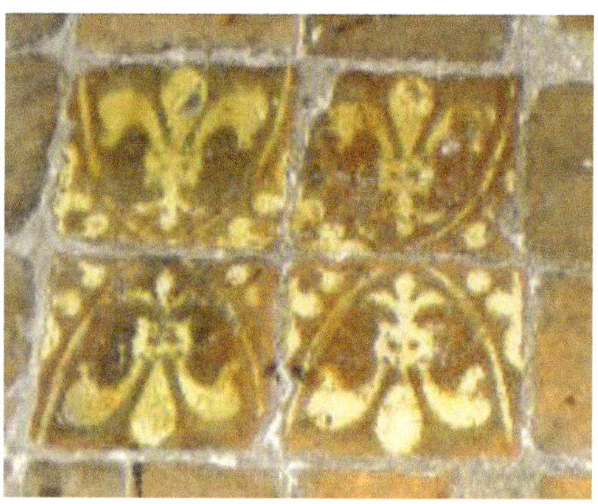	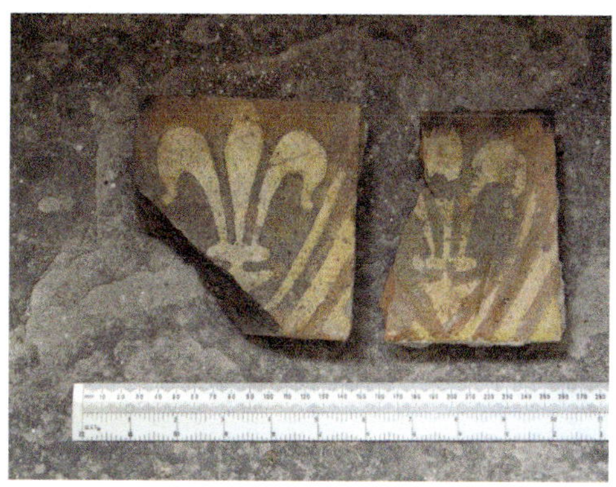 Similar at Wigmore Abbey but with different design at corners and different base to flower.

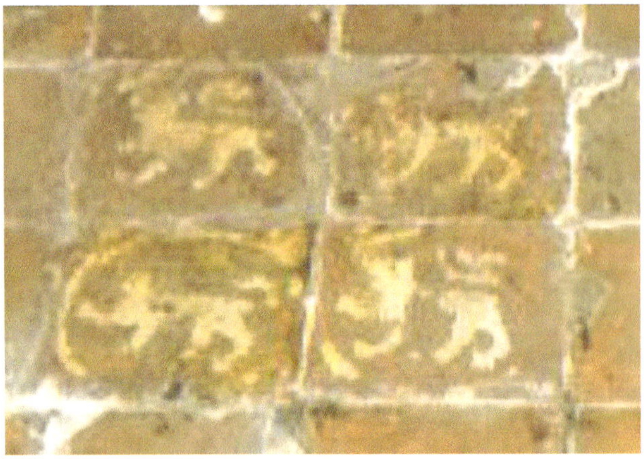	 JM Lewis Group 8 Lion Passant within circle with fleur de lys at corners. BM Design 1783 (Cheltenham) Monmouth Netherwood, Thornbury Wigmore Abbey Worcester Cathedral Goodrich Castle (crude version).
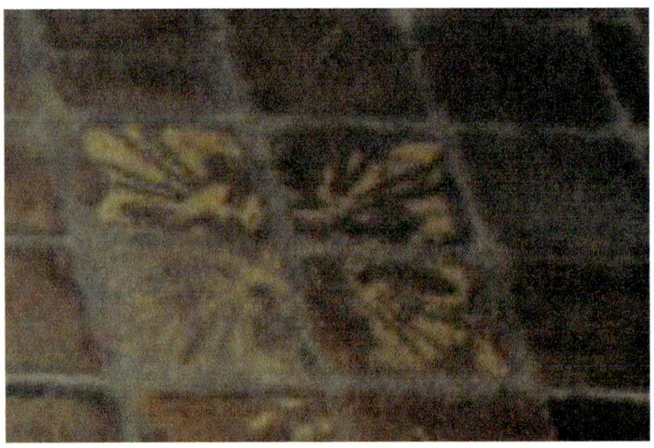	 Similar at Netherwood, Thornbury Similar at Worcester Cathedral.
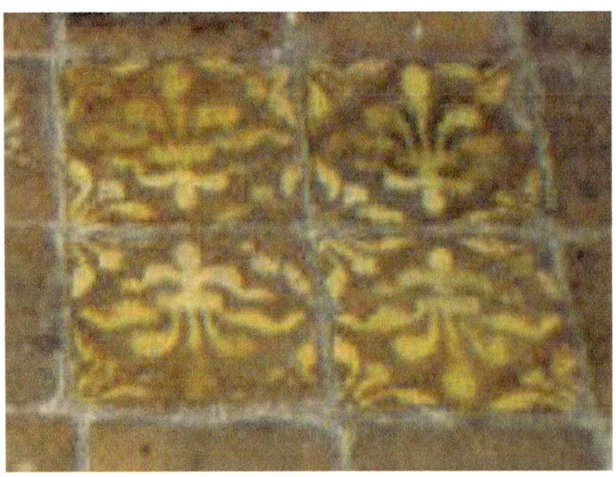 Unknown.	

Hereford – Eign Gate

In the Herefordshire Archives (D69) are a collection of drawings by a local antiquary, Thomas Bird, produced circa 1828. Amongst these are drawings of ten medieval floortiles with the accompanying description of their find site.

'As some men were lately employed in sinking a cellar on Mr Huxley's premises, the south side of Eign Street, Hereford, about nine feet from the surface of the earth, they found a tessellated pavement. The square bricks of which it was composed are many of them ornamented with different devices; many of them have green and black vitrified surfaces, and a portion have the arms of our early Saxon kings, and other arms of ancient families, on them, nearly as fresh as when from the kiln. Amongst others are the arms of Egbert and Ethelbert, a cross forme or; of Edward the Elder, a crosse forme between four martlets or. ..... The pavement must doubtless have belonged to a building of some importance, though no record exists than can lead to a correct conclusion on the subject.'

Thomas Bird's drawings are watercolours which are now rather indistinct and therefore not easy to reproduce here. However, eight are heraldic shields; one is the winged and horned beast from Dore Abbey; and one similar to a Leominster tile with a stylised mouse/rabbit under a vine in a corner of a 4-tile design. The actual tiles are now lost to us.

In order to identify the heraldic shields I have used Tony Wilmott's drawings from the Vince and Wilmott paper on 'A Lost Tile Pavement at Tewkesbury Abbey', together with his descriptions.

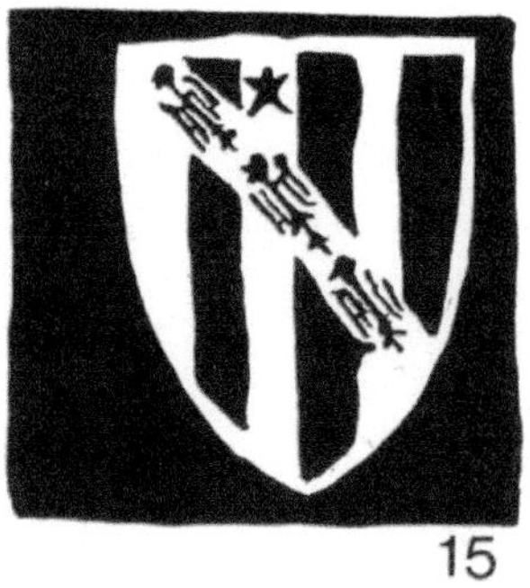 15	*Paly azure and argent.* Possibly William de Grandison (before 1335) Not identical: The Eign Gate tile has a clear diagonal band with no eaglets, and there is no star (mullet) at the top of the third stripe. Also at Bredon, and St Oswalds, Gloucester.
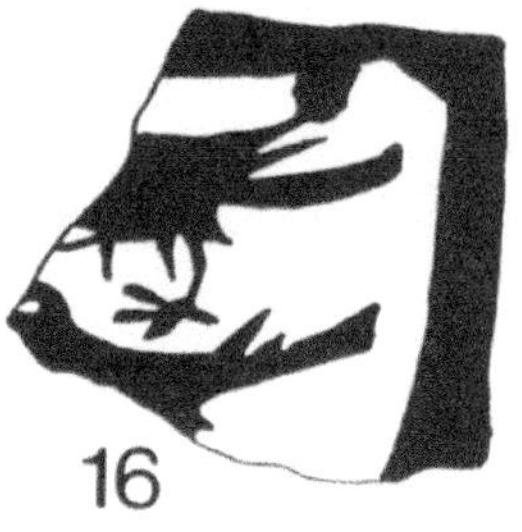 16	*Or, three corbies sable.* The Eign Gate tile has three birds. Possibly Morton or Chetinton – branches of the Corbet family (Shropshire). Also at St Peter's Ludlow.

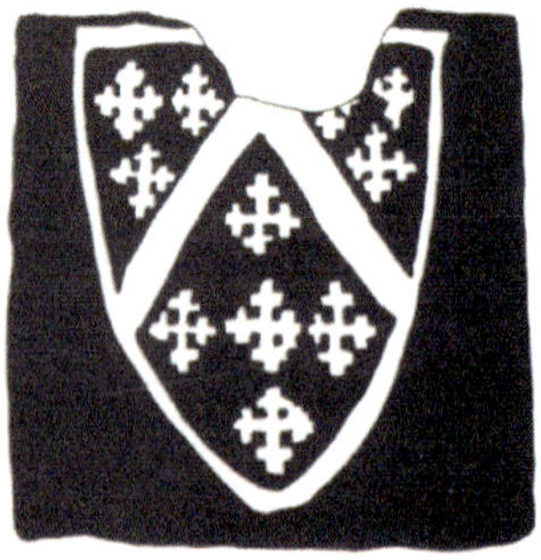 20	*A chevron between nine crosses crosslet* (here there are ten). Often confused with the Berkeley family arms; or de Kyme family. Also at Bredon.
 35	*A cross patonce.* Unidentifiable – very common. Also at Bredon and St Oswalds, Gloucester.
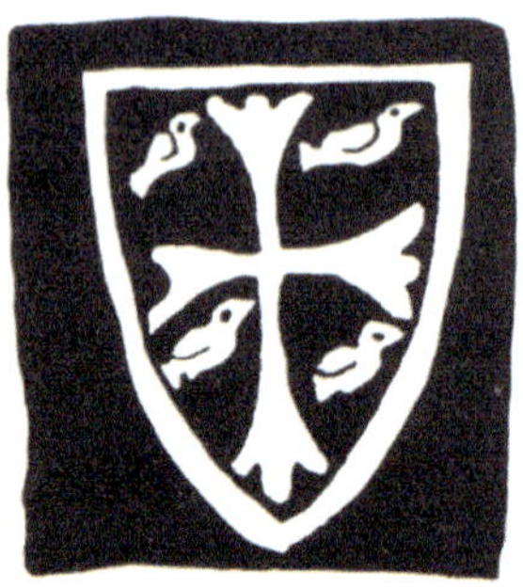 36	*Azure, a cross patonce between five martlets, or.* Edward the Confessor. Based on the coins of the Saxon king who was the patron saint of Kings Edward I-III. Eign Gate has only four martlets. Also at Dore Abbey.
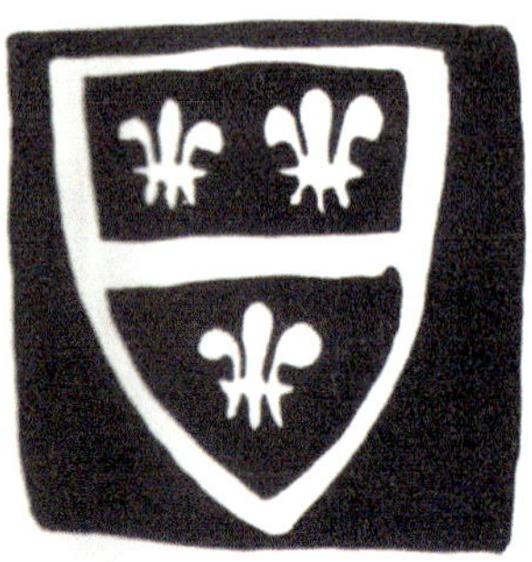 43	*A fess between three fleurs de lis.* Likely to be Edvin Ralph of Herefordshire. Also at Dore Abbey.

 50	*A maunch.* Possibly the Hastings family, based at Abergavenny who were part of the Mortimer faction. Or Robert de Tony based at Painscastle. Also at Bredon.
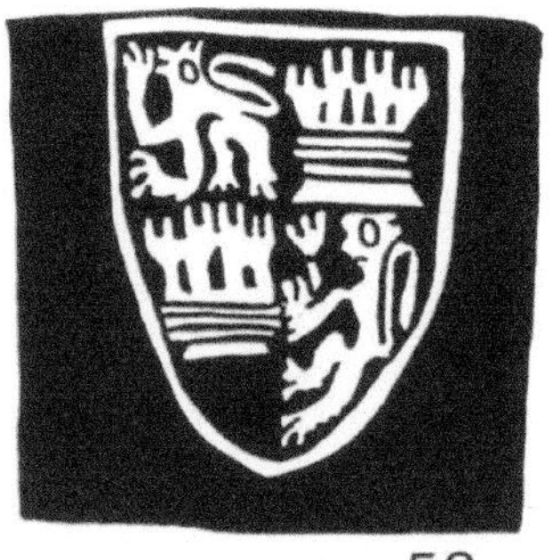 53	*Quarterly, 1 and 4, gules, a castle or; 2 and 3, argent, a lion rampant purpure.* Arms of the Spanish House of Leon and Castile (after 1254) (NB: The quarterings are represented on the tile stamp in the wrong order.) Also at Dore Abbey, Bredon, and St Oswald's Gloucester.
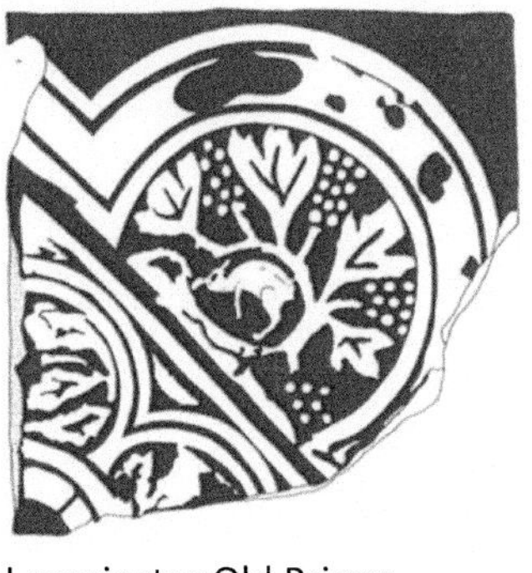 GS10 from Leominster Old Priory	This tile from Leominster Old Priory is the same design as the one from Eign Gate. Each corner has an exaggerated semi-circle featuring a rabbit or mouse beneath a stylised grapevine. The straight lines would have produced a diamond-shaped central motif within the 4-tile group. Chertsey/Halesowen BM 2721 depicts the same design but with superior workmanship.
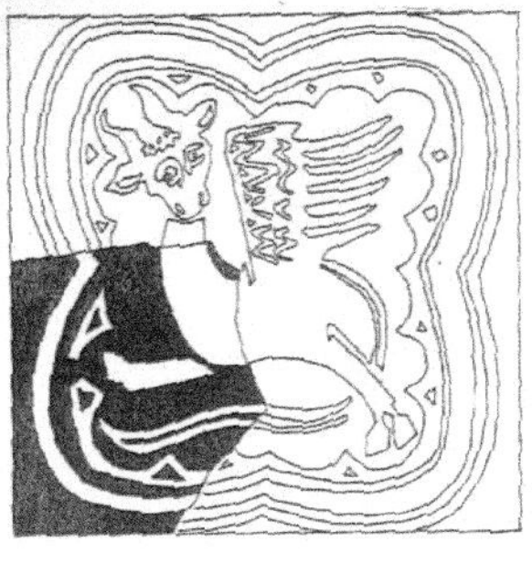	This winged and horned beast is also found at Dore Abbey and Blackfriars, Gloucester. The design matches BM 1389.

Hereford Museum Collections

Fragments of medieval floortile are curated in the collections from many excavations carried out during redevelopments within the town. The main sites are:

- Berrington Street
- Bewell Square
- Blackfriars
- Breinton
- Brewery site
- Bus Station (St Guthlacs)
- Canal Road
- Castle Street
- CRS Store
- Deens Court
- Garrick House
- Greyfriars
- 28 Greyfriars Avenue
- Hereford Castle
- Hereford Cathedral
- Mappa Mundi
- Maylord Street
- Tesco
- Trinity Almshouse
- Wall Street
- Widemarsh Street

The majority are too fragmentary to recognise any design which might be matched to known examples from elsewhere. I include a few of the more interesting ones, listed alphabetically.

Berrington Street

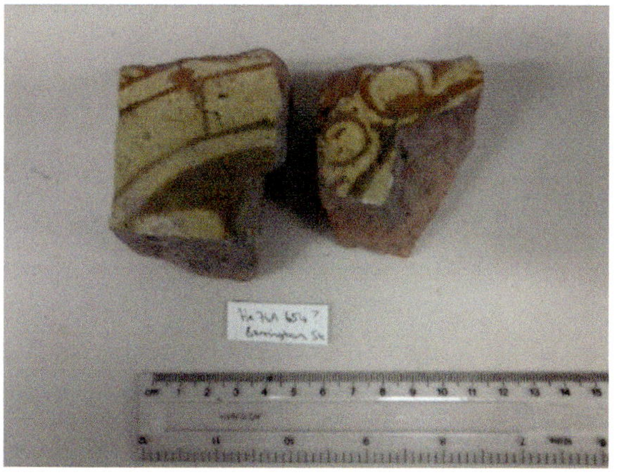

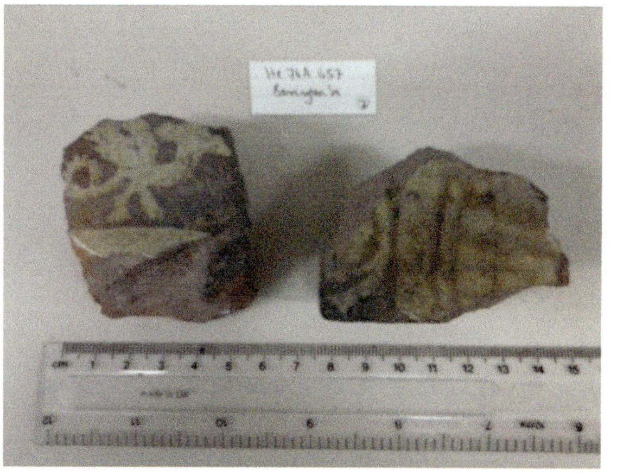

This fantastical beast is similar to a tile found at Leominster Old Priory (GS13).

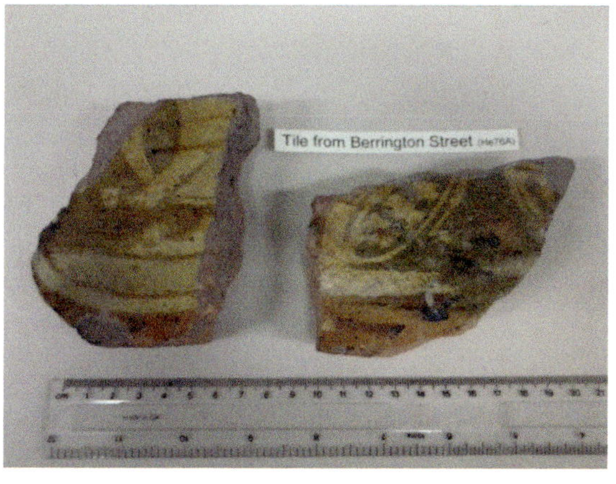

The right hand fragment is probably this design:

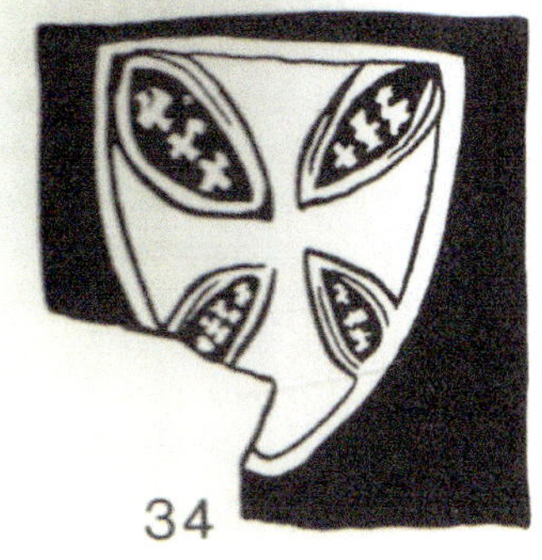

34

Berenger Moyne

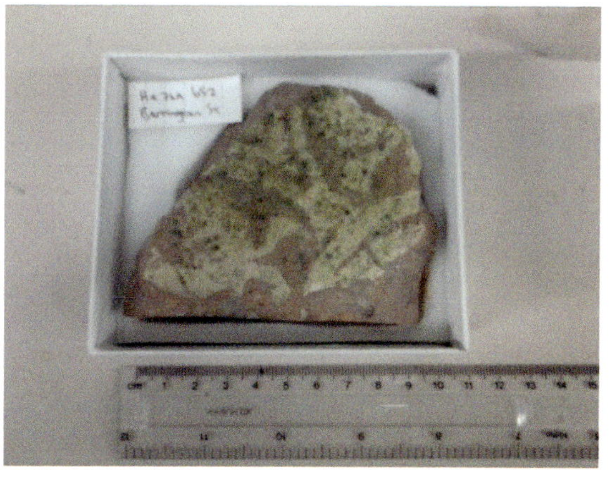

Blackfriars

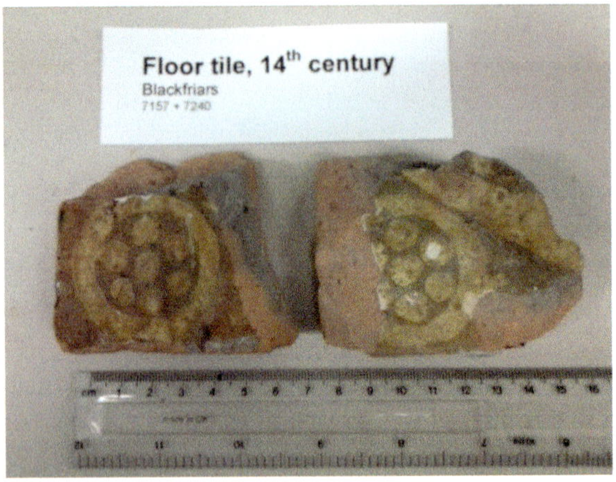

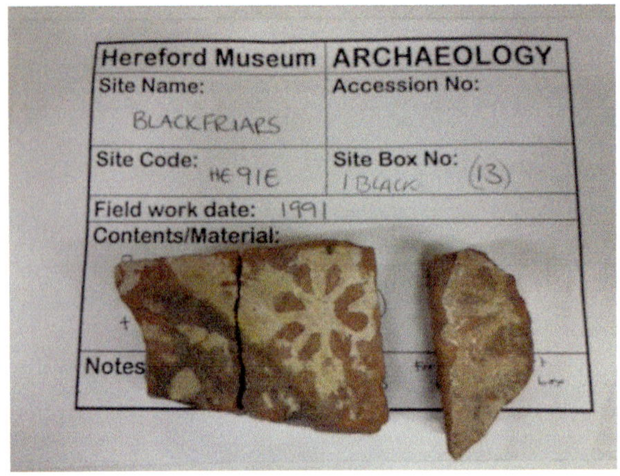

Breinton

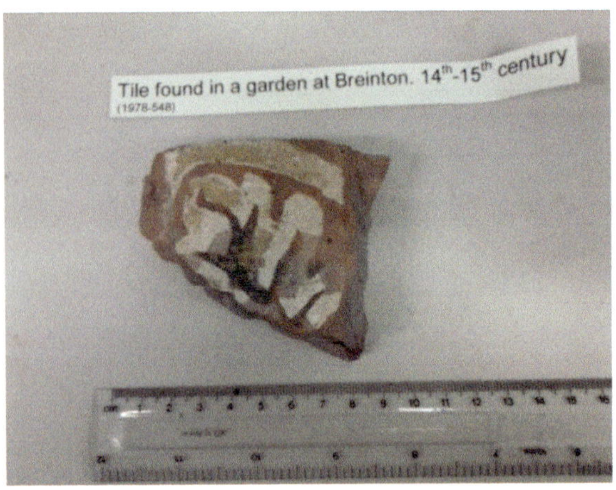

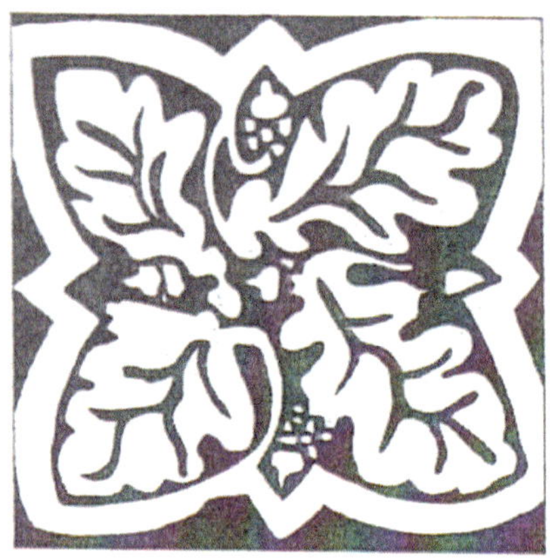

This Breinton tile would probably have formed part of a design similar to this from Much Marcle.

Canal Road

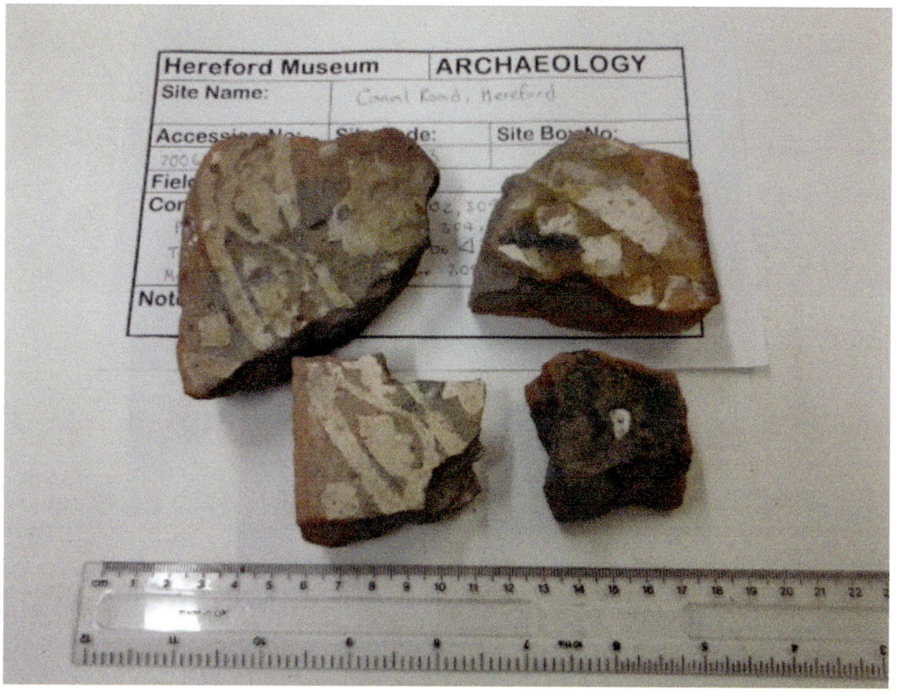

Castle Street

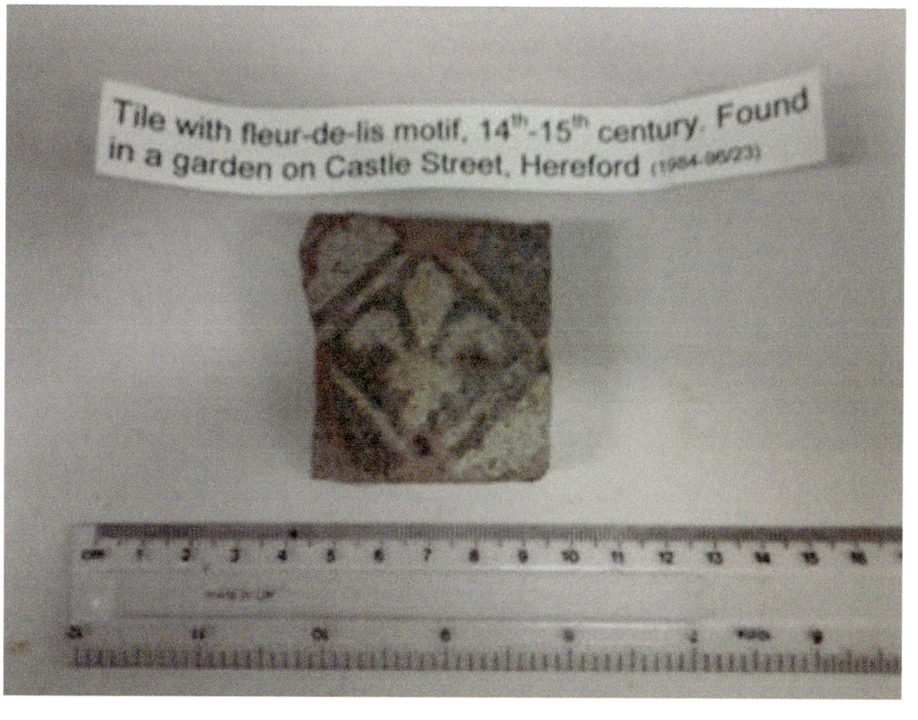

Deens Court

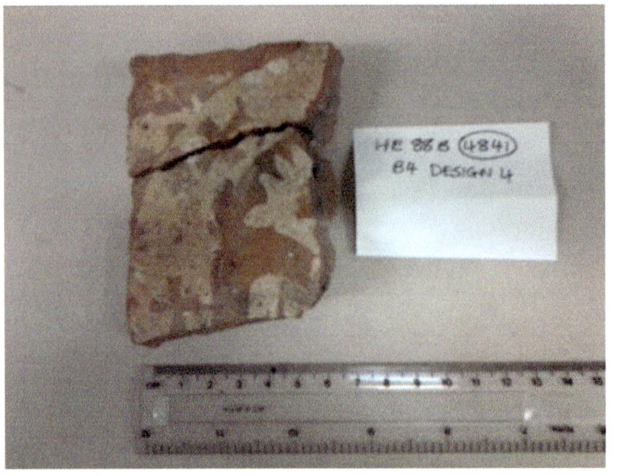

One of the tiles from Dilwyn has a similar image of a spikey headed creature.

Greyfriars

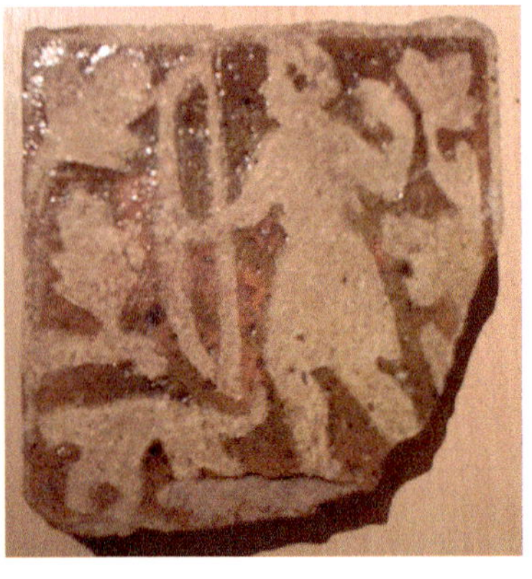

Also at Wigmore Abbey

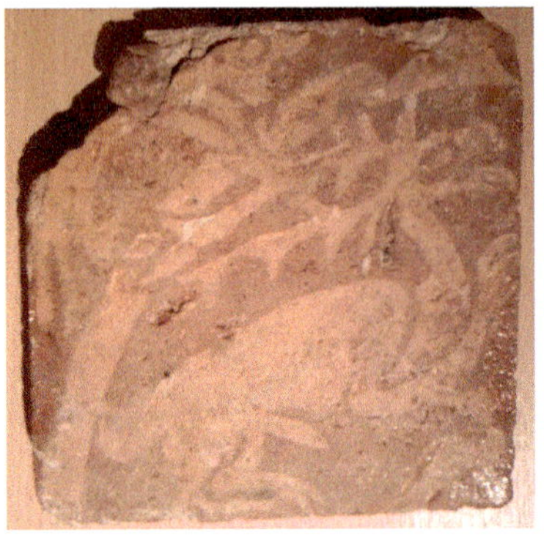

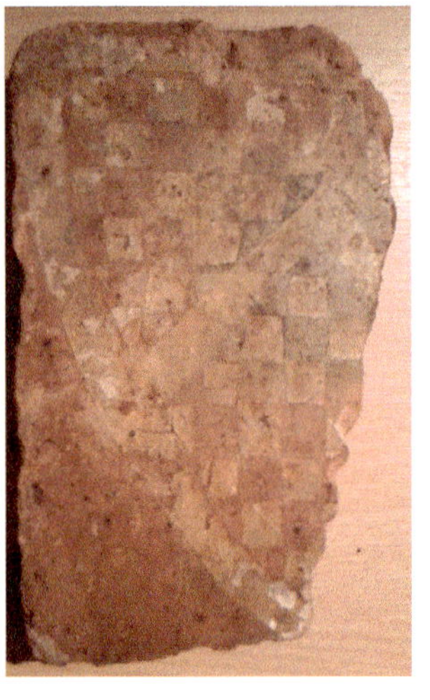

Chequy, a chevron
Boterels, de Arderne, or Newburgh?

Trinity Almshouse

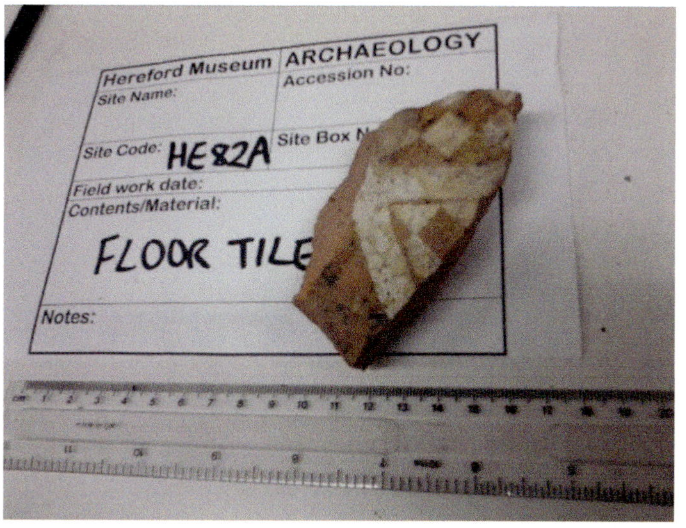

Wall Street

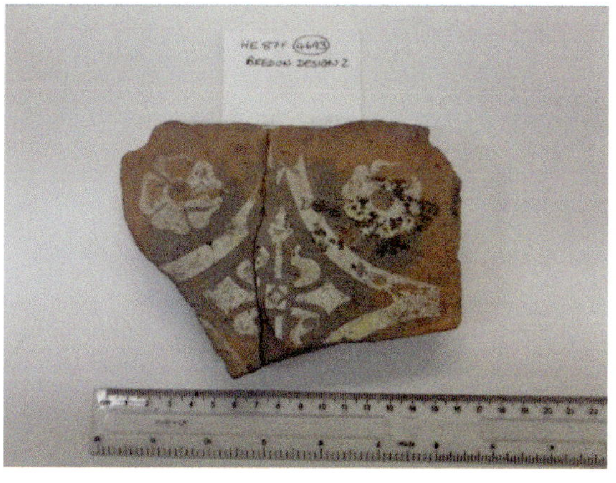 Different floral motifs.	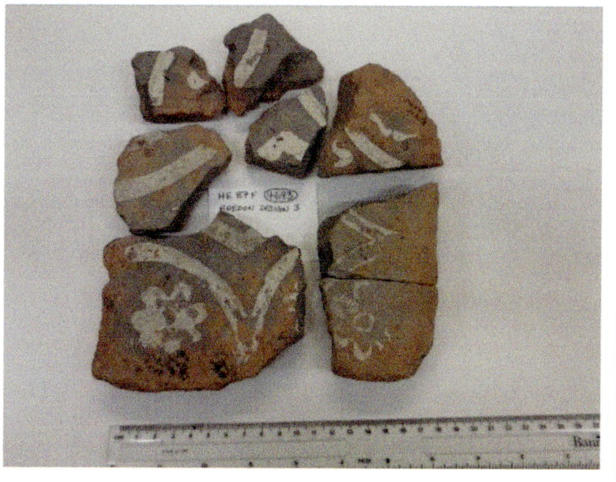

Ledbury – St Katherine's Chapel

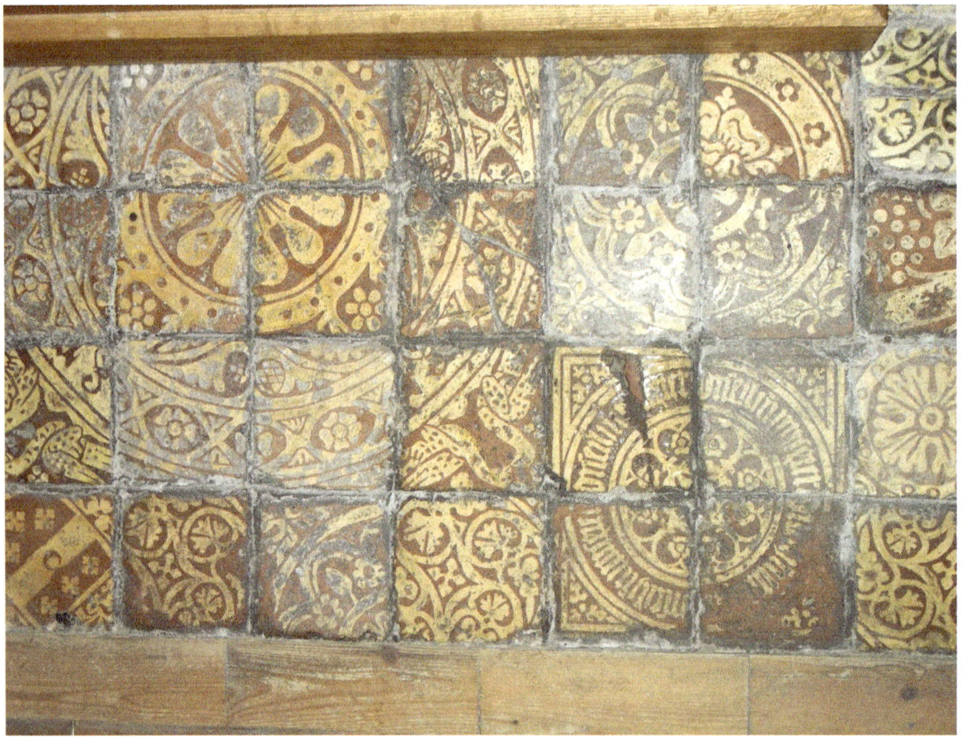

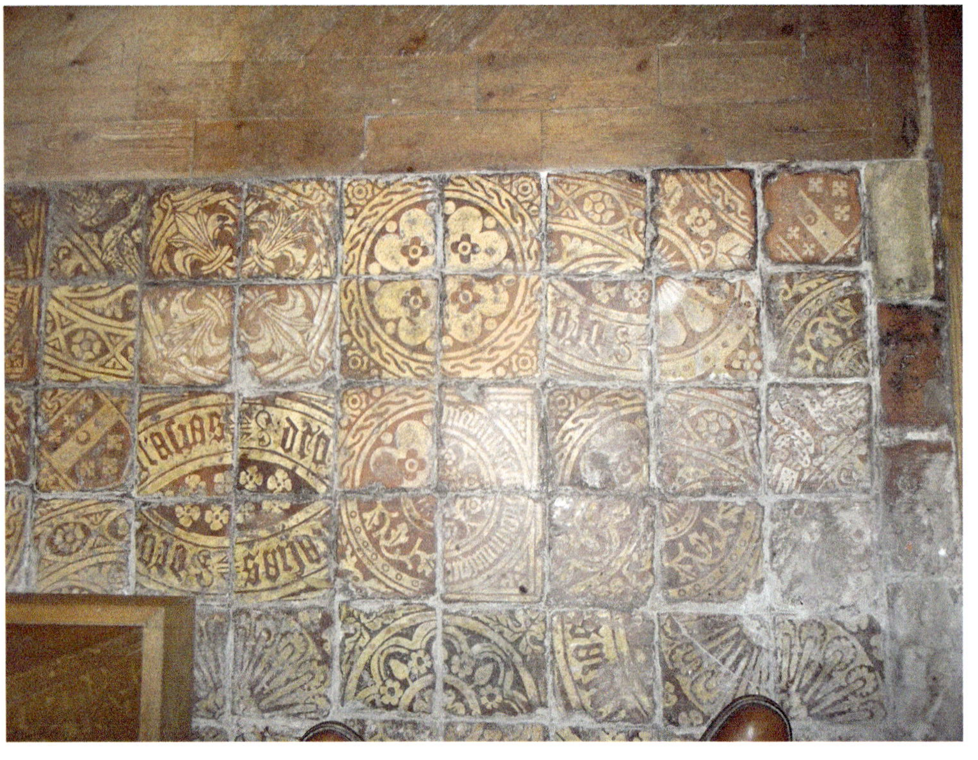

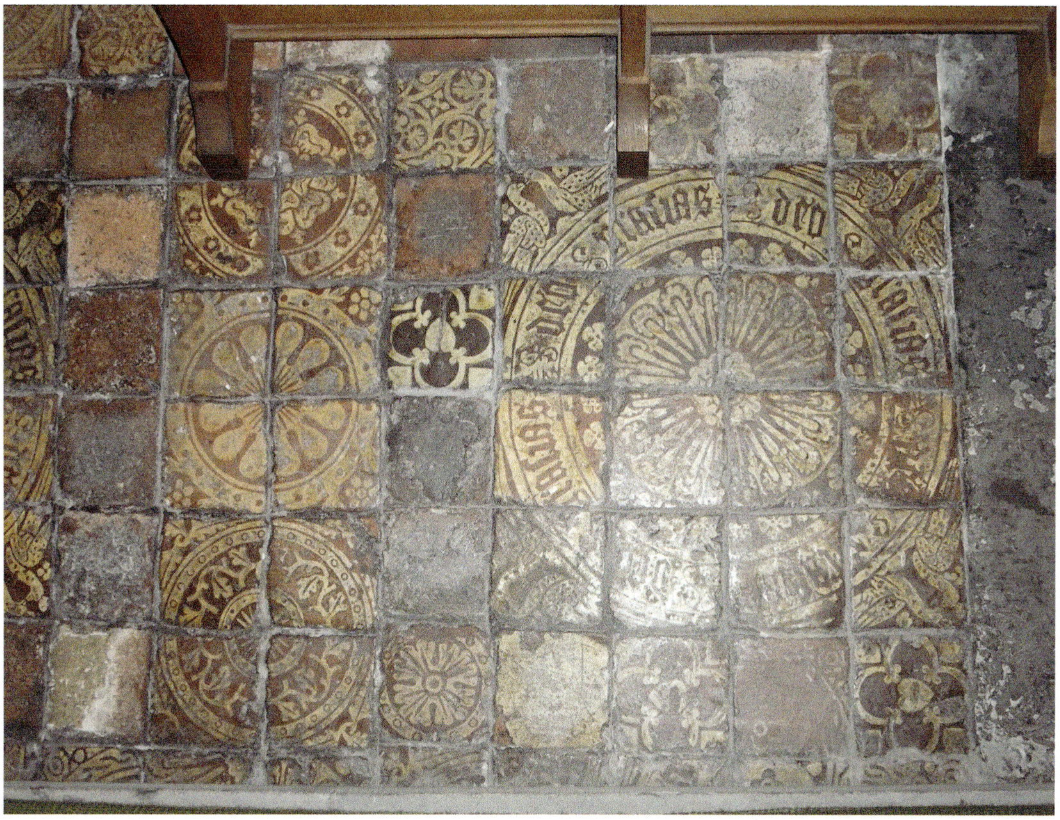

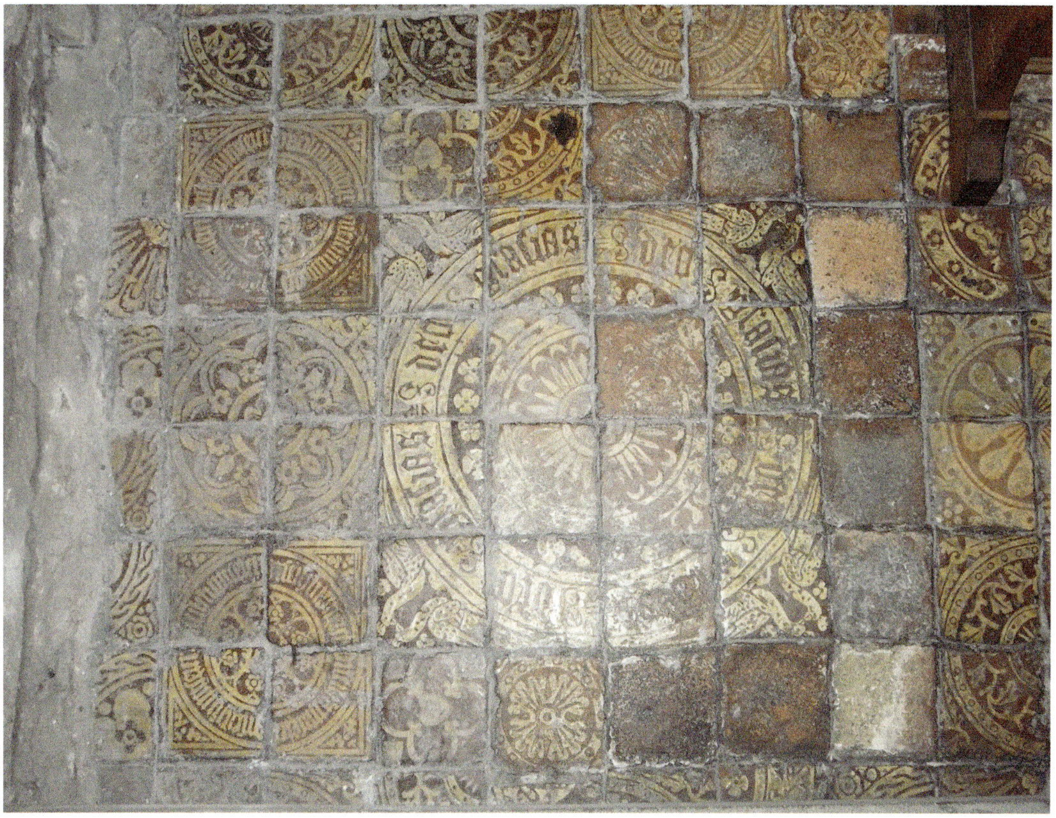

This is such a delight to visit! Such a treasure to have survived ... Joe Hillaby has written an excellent book on St Katherine's at Ledbury in which there is an extensive chapter on the chapel's late medieval tiled pavement. I would refer any reader to this publication (St Katherine's Hospital at Ledbury, 2003). Access to the building is by prior appointment.

I will only give a brief summary here. There are around 500 tiles (with an approximate date of 1485), the vast majority of which are of the Canynges group. There are examples of six different sixteen tile patterns, identifiable from the wide exterior banding. There are four examples where four tiles clearly form a circle, again identified by the smaller circular band; and a further six stand-alone designs which could equally have been grouped in a four.

I will identify them here by using the JM Lewis drawings and descriptions. He lists them in his book as belonging to Group 24 and gives an approximate date of c.1480-1530.

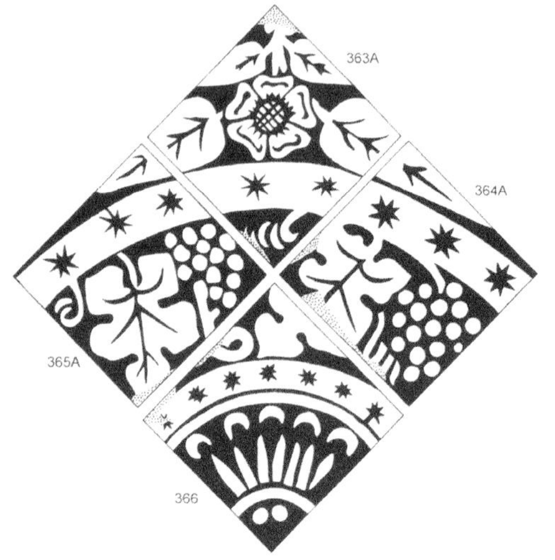	363 16 tile design; circular band with reserved stars surrounding a frieze of vine leaves and bunches of grapes round an inner starred band surrounding a sun-burst; in each corner a five-petalled rose with leaves. BM Canynges Design VI. Also at Shelsley Walsh, Worcs.
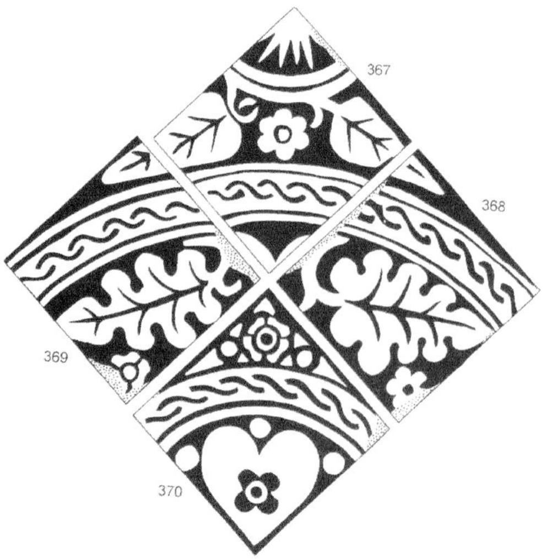	367 16 tile design; circular band with simple cable pattern enclosing oak leaves; the circular band is repeated at the centre, where it encloses a heart pierced with a four-petalled rosettes; in each corner outside the larger circle a six-petalled rosette flanked by leaves springing from a double circle surrounding a sun-burst. BM Canynges Design I. Also at Deans Place, Yatton.

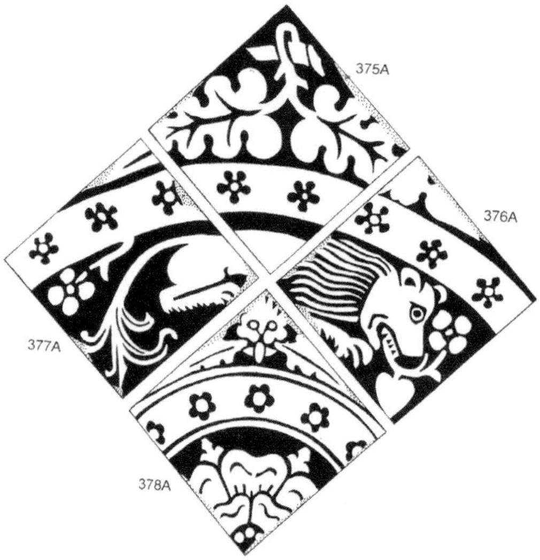

375

16-tile design; circular band of pierced cinquefoils surrounding a frieze of crouching lions; a similar inner circle surrounds a double eight-petalled rose; in each inner spandrel is a lion mask with leaves issuing from the mouth, and in each outer spandrel two oak-type leaves on a stalk.

Also at Shelsley Walsh, Worcs; Blackfriars, Gloucester.

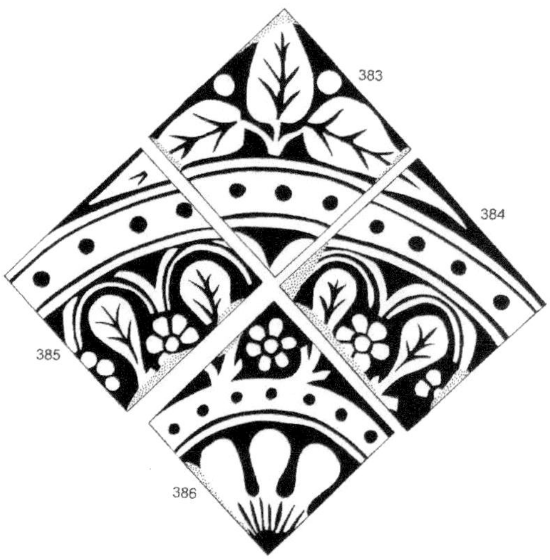

383
16-tile design; circular band with reserved pellets surrounding a frieze of arcading, each arch containing a leaf that springs from an inner circle (also with pellets), between each pair of leaves a siz-petalled rosette; in the centre a twelve-petalled flower; in each corner outside the circle three lanceolate leaves with pellets between.

BM Canynges Design VII.

Also at Deans Place, Yatton.

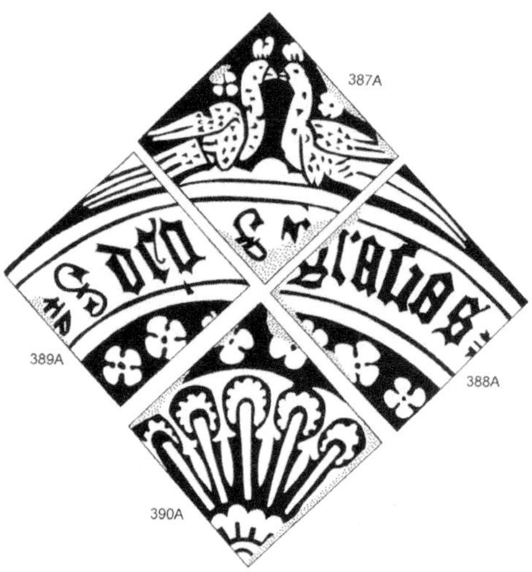

387
16-tile design; Circular band with reserved black-letter inscription: *deo/gratias* repeated four times, surrounding a band of four-petalled rosettes which in turn surrounds a sun burst; outside the circle each corner has a pair of peacocks.

BM Canynges Design II.

Also at Shelsley Walsh, Worcs.

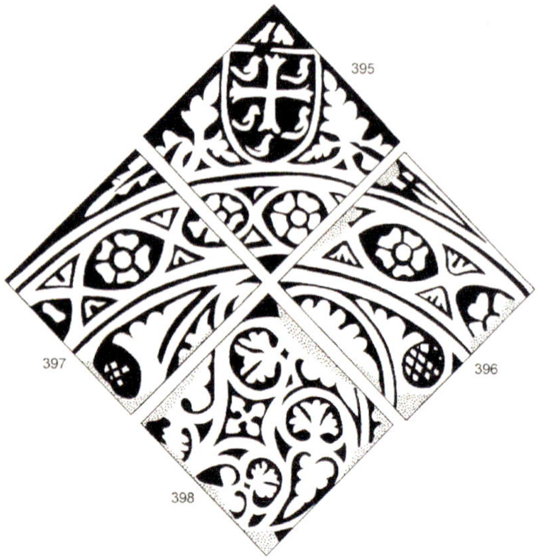

395

16-tile design; circular band with five-petalled rosettes in a cable pattern surrounding a quatrefoil containing an asymmetrical foliage design; in each corner outside the circle a shield of arms: *between five martlets contournee a cross patonce* (Westminster Abbey reversed).

BM Canynges Design IX.

Also at Deans's Place, Yatton.

Four-tile designs:

399A

399 A

4-tile design; circular band with reserved annulets surrounding a leaf scroll pattern; at the centre a circular band with reserved pellets surrounding a sun-burst; in each corner a three-leaf motif.

Also at Shelsley Walsh, Worcs.

402

4-tile design; circular band of arcading; with six-petalled rosettes at the junctions; under each arch a simple leaf springing from a central ring; eight petalled flower at the centre; in each outer corner a three-leaf motif.

BM Canynges Design XVII (2796).

Also at Shelsley Walsh, Worcs.

407A

407

4-tile design; four fleur de lys set diagonally, their tips breaking a surrounding double circle; in each outer corner a six-petalled rosette and intertwined leaf motif.

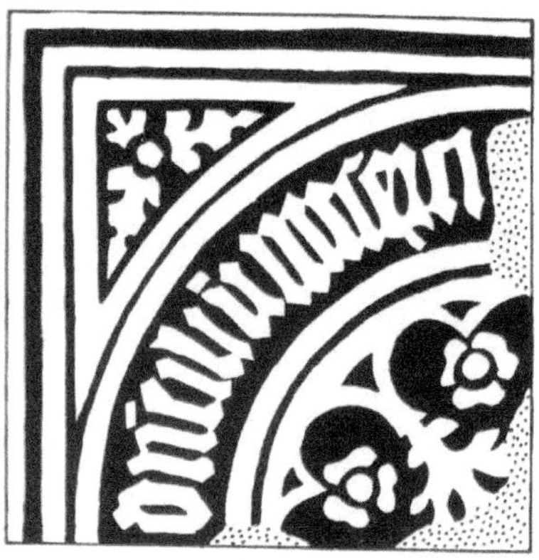

416

416

4-tile design; circular band with black-letter inscription *dne ihu miserere* (Lord Jesus have mercy) repeated four times; within the circle, tracery with three-petalled rosettes at the angles; in each outer corner a leaf motif in a frame; the whole enclosed in a double-line frame.

BM Canynges Design XI (1442).

Stand alone tiles:

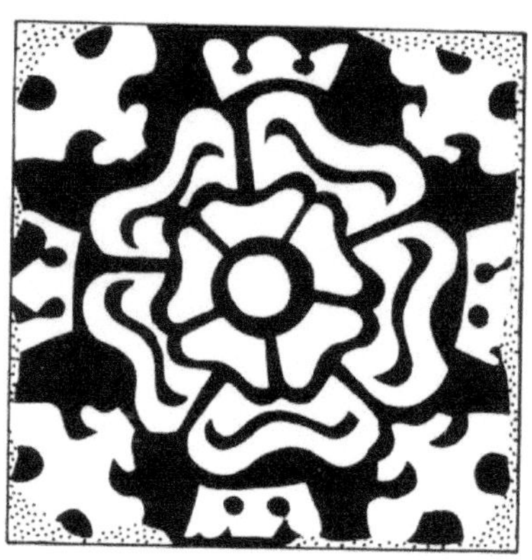

412

412

Slip on impression. 'Tudor' rose crowned on four sides; a fleur de lys motif in each corner.

Also at Shelsley Walsh.

 413	413 Slip on impression. Quatrefoil with fleur de lys springing from the angles inside and out. BM Canynges Design XXI (2204).
 414	414 Slip on impression. Quatrefoil with trefoils at inner angles, enclosing a heart, and with a heart set diagonally in each outer angle. BM Canynges Design XII (2008).
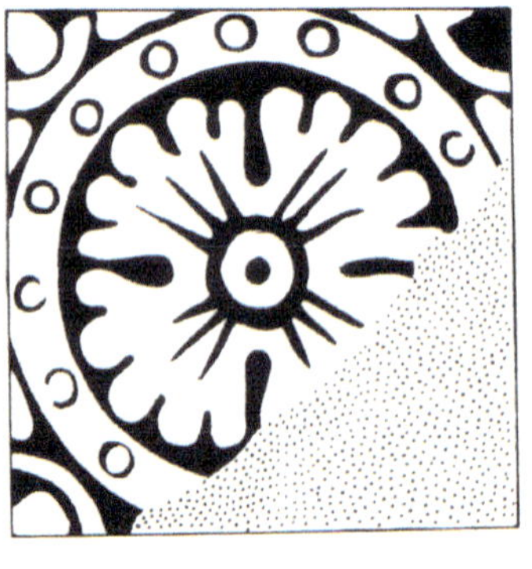 415	415 Slip on impression. Continuous pattern, but used in 4-tile blocks; circular band with annulets enclosing a flower motif of four petals with convoluted edges; quarter circles in each corner. BM Canynges Design XXVII (2366).

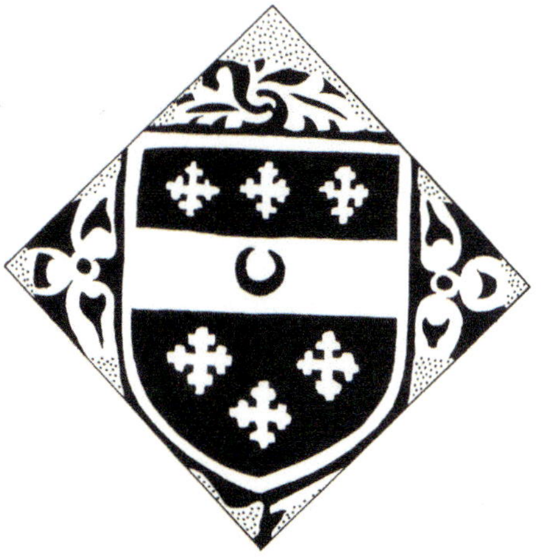 443	443 Slip on impression. Shield of arms, set diagonally, *on a fess between six cross crosslets a crescent* (Beauchamp, ?differenced for William, 1st Lord of Abergavenny); leaf motifs above and at sides. BM Canynges Design XVI (1606).
 320	The only non-Canynges Design: 320 Slip on impression. 4-tile design; within a dircle four shields of arms set diagonally each within a ring of stylized foliage; *a fess between six cross crosslets* (Beauchamp); outside the circle each corner has as stylized foliage motif. BM Design 1601 (Great Malverrn Priory, Lenton Priory kiln site). Also at Stretton Sugwas, Monmouth, and Raglan.

Leominster Old Priory

Leominster has been the site of a religious settlement for 1300 years, with the foundation of a church here by St Edfrith in 660. The church that is visible today first appeared in 1123 when King Henry I founded a Benedictine Abbey at Reading, granting it the income from Leominster. Leominster and Reading were both subject to the Dissolution in 1539, and the floortiles found are likely to date to this destruction.

In a small recess at the western end of the south nave are some fragments of medieval floortile salvaged during George Gilbert Scott's restoration work on the priory church in the late 19th century.

In the early 1990s Hilary White and Duncan Brown published drawings, together with a brief interim statement on these tiles, and this information was incorporated into Joe and Caroline Hillaby's book on Leominster Minster, Priory and Borough (2006, 192-6). Duncan Brown holds the copyright for all the drawings of these tiles. I have numbered these fragments with the prefix of GS (Gilbert Scott).

In 2002 Border Archaeology carried out a watching brief in the south-east corner at Leominster Old Priory which recorded several 'loose-laid' fragments of decorated medieval floortile. No drawings were published but written records give details of 'heraldic designs, flowery and fleur-de-lys decorations' and 'a chevron and line border surrounding worn yellow dots' plus a 'flower design within a double circle border'. There is also a description of an 'upright man raising the palm of his hand behind a lion' – an image I am not aware of (SMR:31935 NGR SO 4980 5935). I have been unable to track down the actual tiles.

An excavation was carried out in 2005 led by Bruce Watson of MOLAS which revealed a large quantity of broken fragments of medieval floortile, on which I reported in 2007. I numbered these fragments with the prefix BW (Bruce Watson); the drawings are by the author.

The tiles below are to be found at the western end of the south nave (drawings by Duncan Brown).

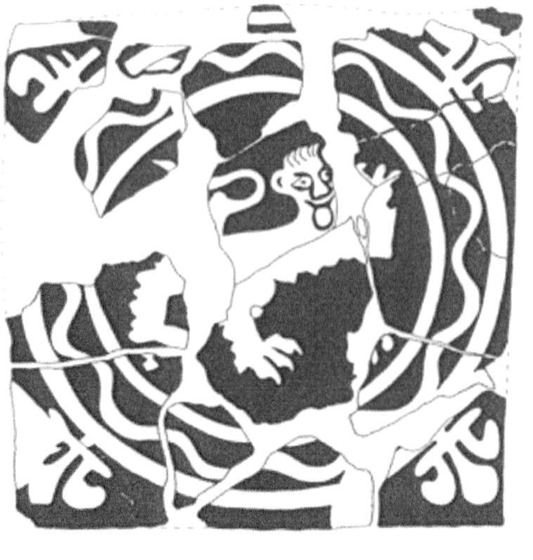 GS14	A lion passant guardant within a double circle containing a wavy line with a fleur-de-lys at each corner. No parallels.

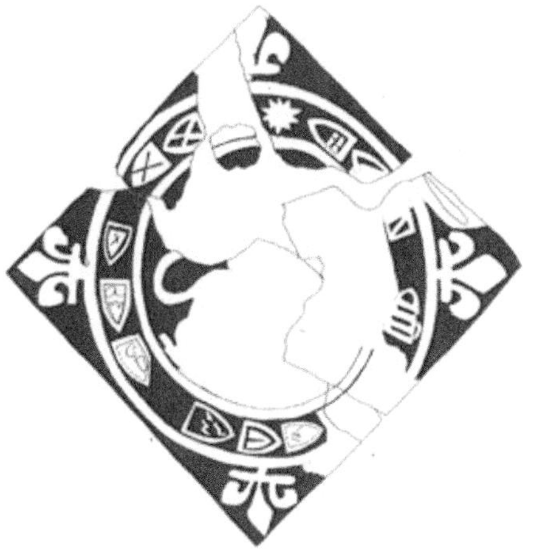

GS16

Tile **GS16** is a self-contained design continuing the circular theme of the lion tile above, and with similar fleur de lys in the corners. Small heraldic shields replace the wavy line encircling a central figure,which was probably a lion, suggested by the loop of the tail.
No parallels.

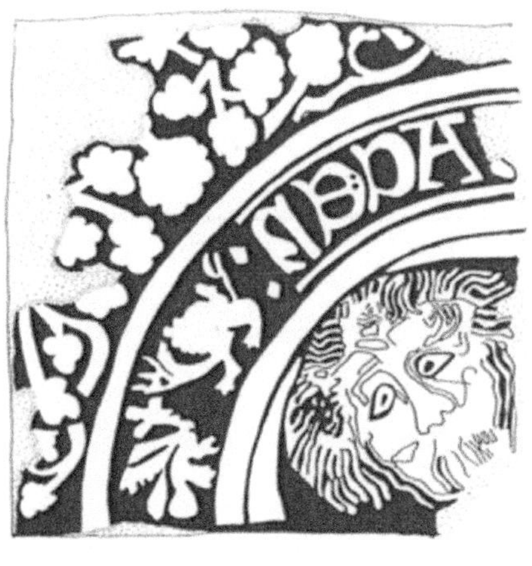

GS12

GS12 can be recognised in BM 1036 (Broadway Priory 15/16th C) and BM 1033 (from an unknown site in Worcs 15/16th C) but these do not include text.
The half human/half animal face is surrounded by bristling hair, encircled by fantastical beasts and Lombardic text. The Leominster workmanship in inferior.

Also at:
Hailes Abbey (Hillaby)
Abbey Dore.

GS13

GS13 has a very close parallel in BM 1035 (Broadway Priory 15/16th C) with virtually identical beasts. The quality of the foliage however is inferior at Leominster.

Also at:
Berrington St. Hereford (Hillaby).

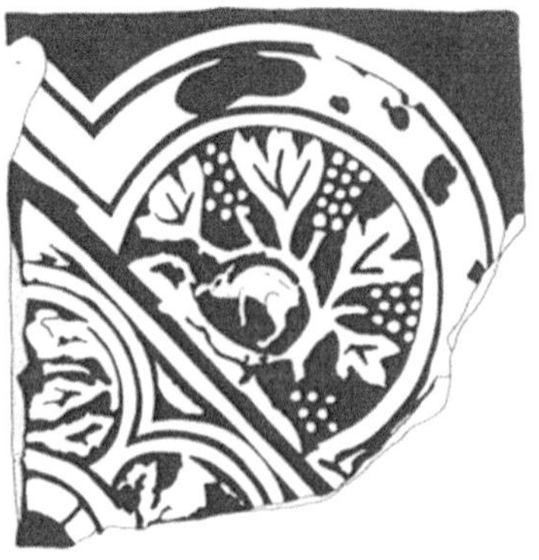

GS10

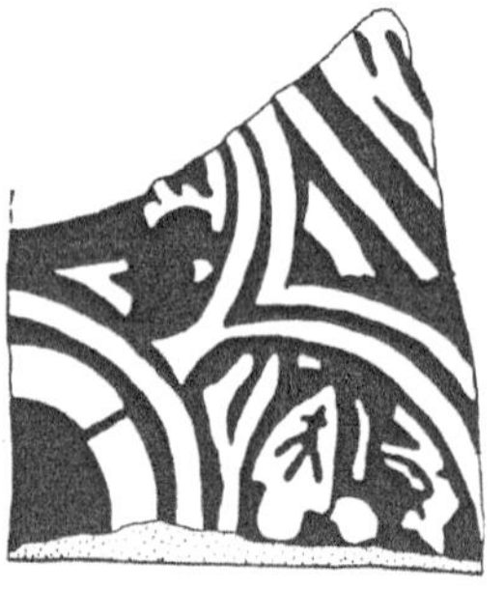

GS38 (same design)

Another pair of tiles, **GS10** and **GS11**, follow the Chertsey/ Halesowen tradition. Here the roundels are replaced by straight lines to produce a diamond shaped central motif. Each corner has an exaggerated semi-circle, one of which features a rabbit beneath a stylised tree or grapevine.
A similar scene appears on BM 2721, but again the Leominster version is inferior.

Also at:
Tewkesbury (Hillaby)
Hailes (Brown)
Blackfriars, Gloucester
Eign Gate, Hereford.

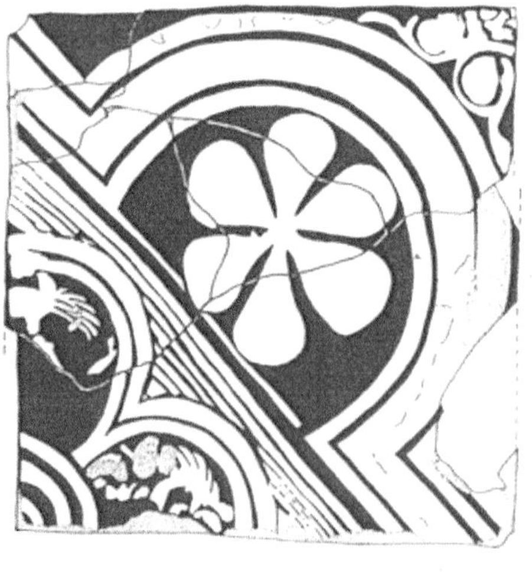

GS11

GS11
As GS 10 above but with six petalled flower in the corner.

Also at:
Tewkesbury (Hillaby)
Berrington St., Hereford (Hillaby)
Similar in Booth Chapel, Hereford Cathedral.

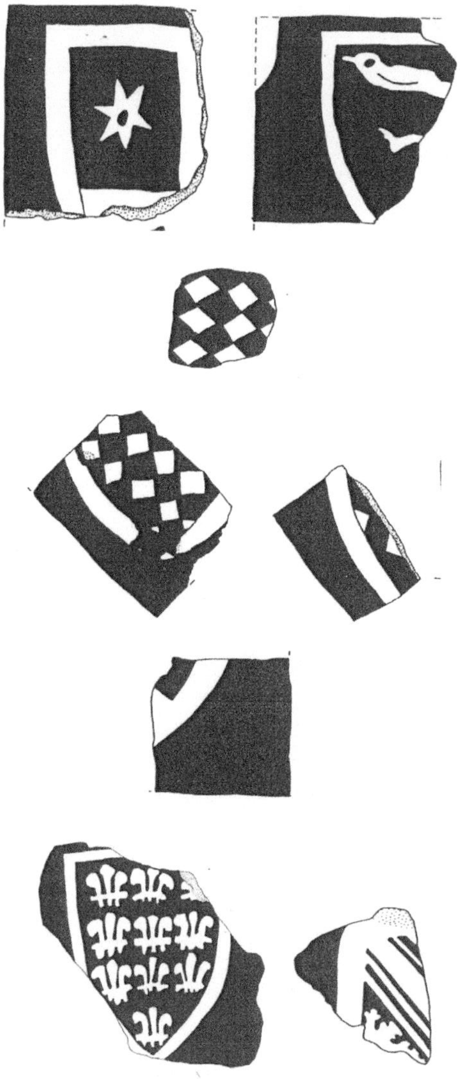

GS25-32

Eight heraldic fragments GS25, GS26, GS27, GS28, GS29, GS30, GS31, and GS32 have parallels with those found in Bredon Church and Tewkesbury *et al* and can be identified with local Marcher lords (Hillaby 2006, 194).

Also at:
Worcester Cathedral and the Singing School, dated to 1370.

GS25 probably formed part of this armorial shield:

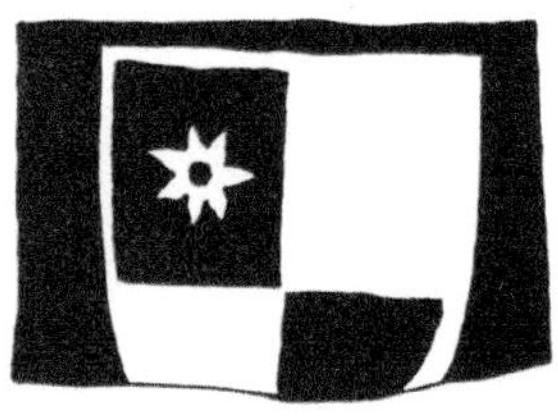

51

Also at Croft
(Probably Robert of Oxford).

Similar tiles of each of these designs have also been found at Worcester Cathedral.

 GS1	 GS2
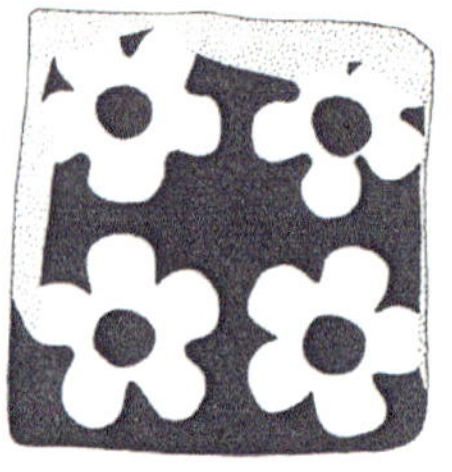 GS3 PH 2096 P084 In Parker-Hore collection 'at Stokesay (also at Leominster)'.	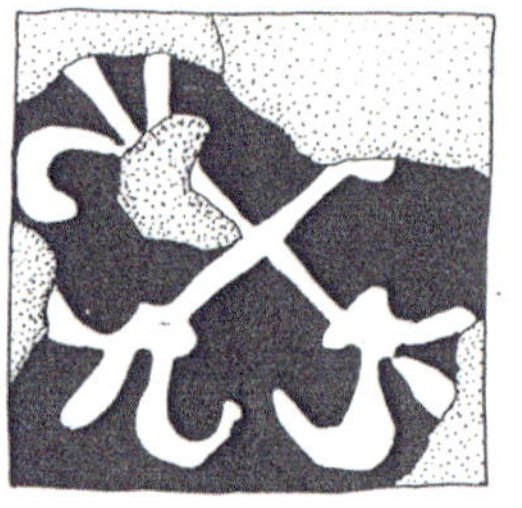 GS8 BM 2186
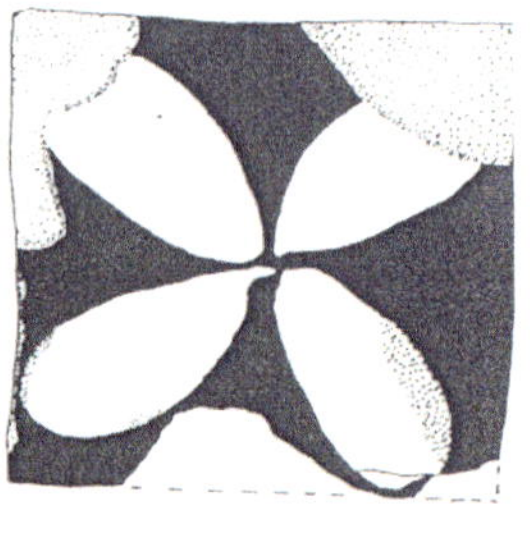 GS4	 GS4 has similarities with this tile from the Parker Hore collection (1067 H525 Leominster) which is also found at Wigmore Abbey.
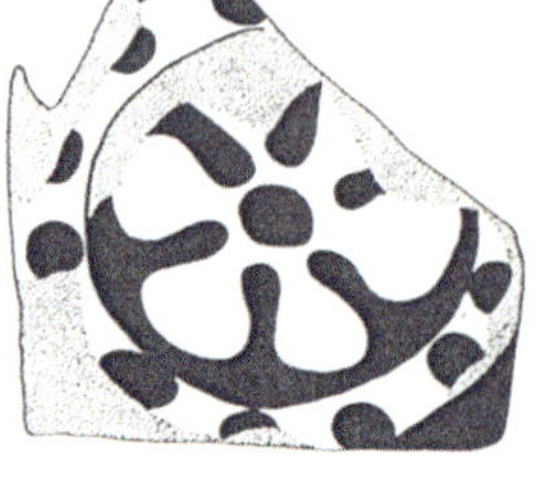 GS5 Similar to BM 2336.	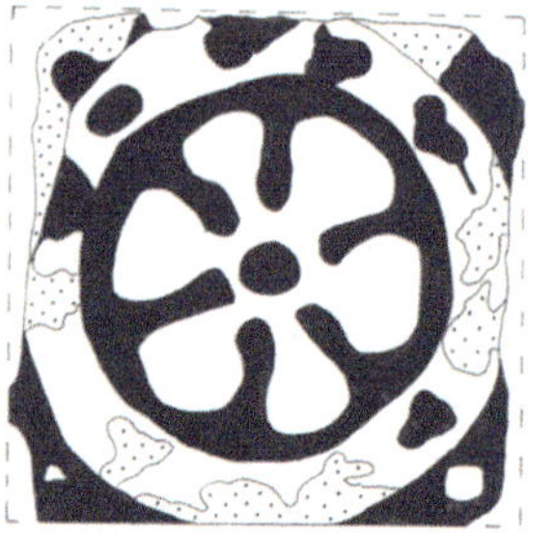 c/f with BW30 Also found at the Commandery, Worcester.

 GS6	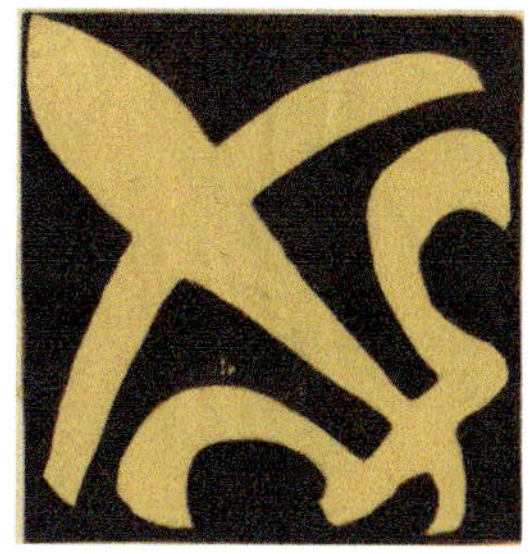 GS6 appears identical to this tile from the Parker Hore Collection (1065 H523 Leominster). Also BM 2243
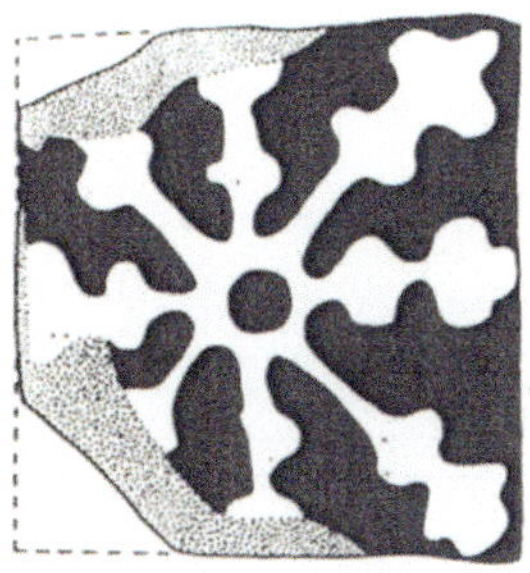 GS9	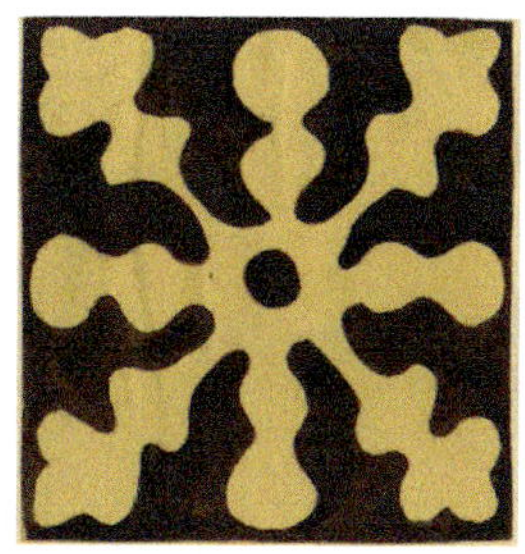 GS9 is also represented here in the Parker Hore Collection (1066 H524 Leominster). Also BM 2479
 GS19  GS16	 2209 GS16 and GS19 would have formed this complete example from Newbridge Abbey, Ireland (14/15 th century). BM 2209

The 2005 Decorated Tiles (from the author's own report 2007)

Leominster Tiles		Parallel Designs
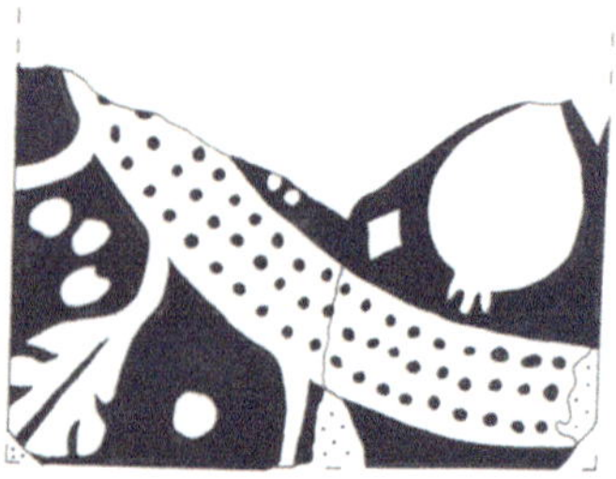	Tiles **BW 1-4** probably formed part of a 16-tile design of the Canynges Group from Bristol. The white circular border with the three rows of small black dots is worthy of note. Laurence Keen suggested it would have been very difficult to create a wooden stamp to produce this design, but with a soft metal such as lead, these small dots could have been easily produced with a blunt spike. A lead stamp would have been mounted on to a wooden board to impress the design. Also at Brampton Bryan. The JM Lewis 383-6 and BM 2987-90 stamps form part of the Canynges Group which share some of the features of this Leominster tile.	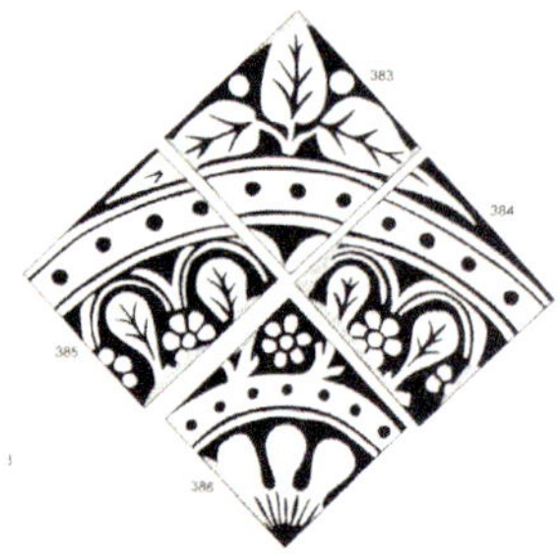 JM Lewis 383-6.

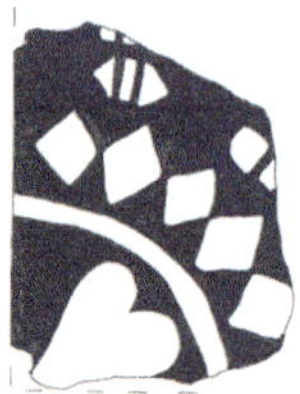 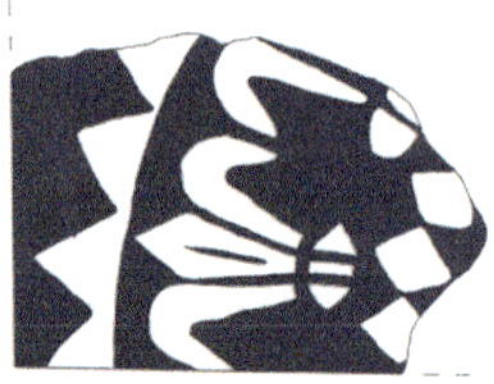 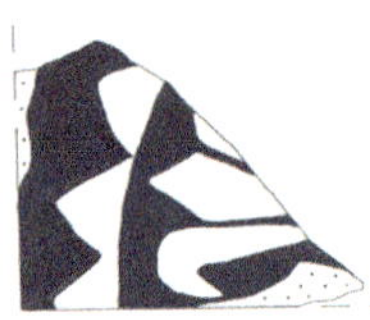 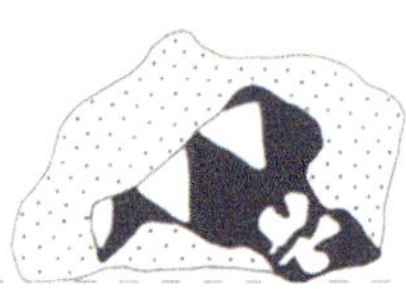 BW8	**BW 5-8** (16 tile design) has a heart motif at its centre. Beyond the heart is a white circular strip which is then separated from a series of fleurs-de-lys by a concentric band of diamond shapes. A chevron border encircles the fleurs-de-lys. Also at Croft, St Michael's and Brampton Bryan (showing a starburst in each outer corner) 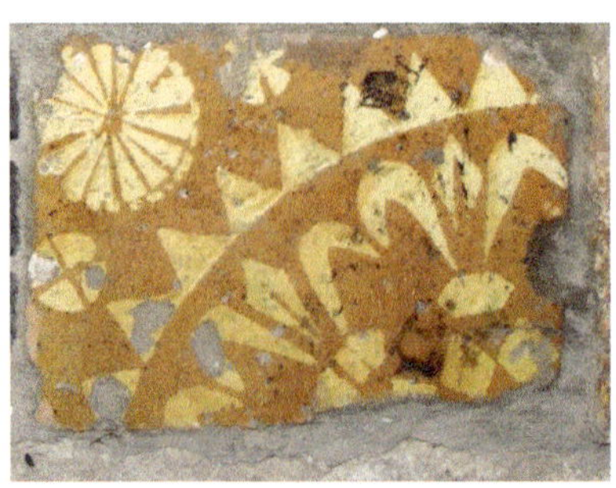 Brampton Bryan	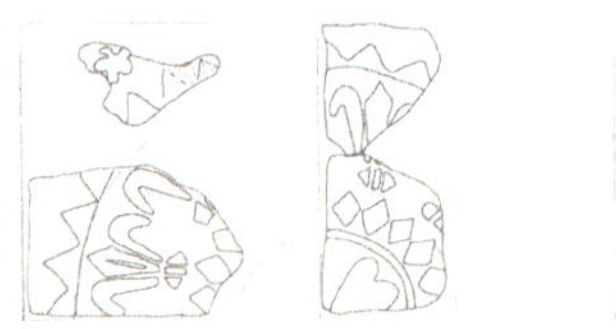 *Drawing of the probable design.*

Leominster Tiles | **Parallel Designs**

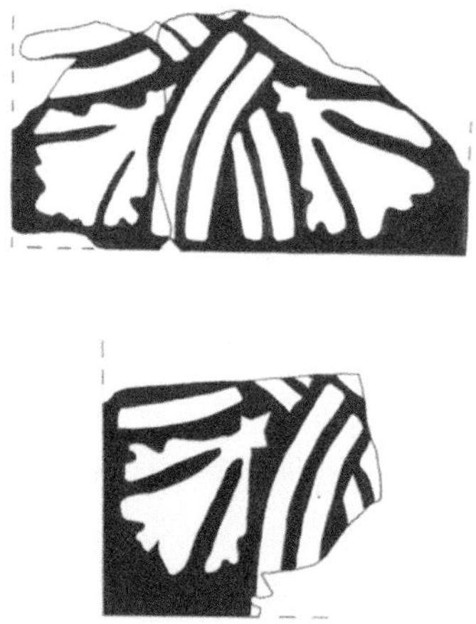

BW 9-11 have a fleur-de-lys in each corner with double interlocking white lines.

Also at Croft, St Michael's and Brampton Bryan.

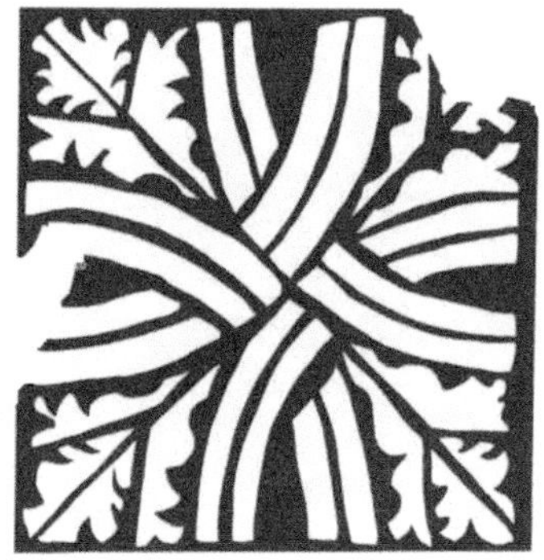

After Eames: BM 2561
Maxstoke Priory, Warwicks. 14th C.

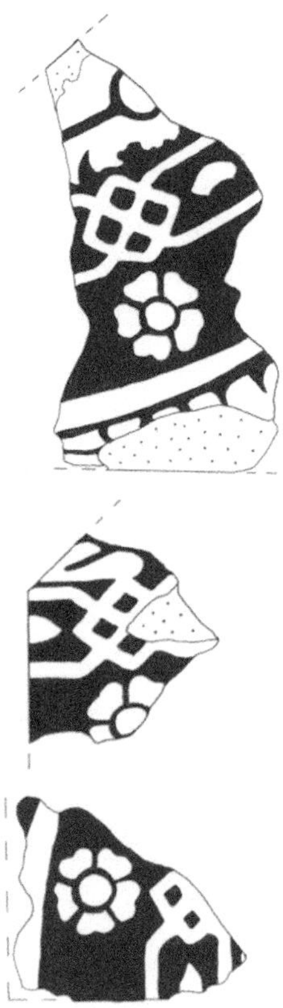

BW 12-19 form the most complex group, as three pieces (12, 13, 16) retain angled sides. I have reconstructed the likely design from the irregular edges found, which shows a central 4-tile insert set at 45 degrees.

This design is identical to the Gilbert Scott tile GS21.

A border of five-petalled Tudor roses is encircled by an outer single white line. Inside this, a concentric design of two widely spaced lines cross over within a square to coincide with the outer Tudor roses. In between the square crossovers there appears to be a lozenge shape within the two lines.

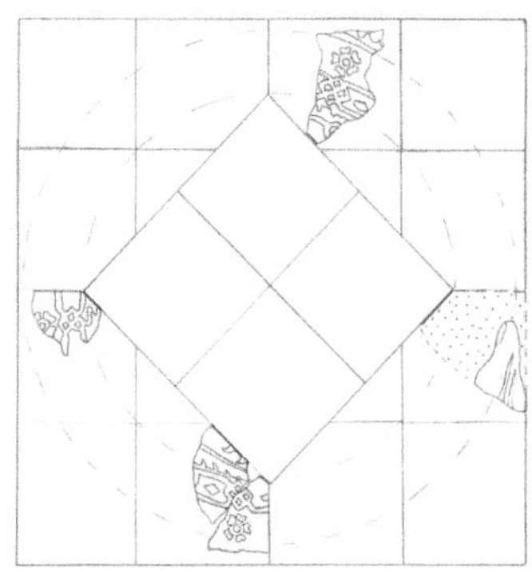

The tile pattern as suggested by the angled sides.

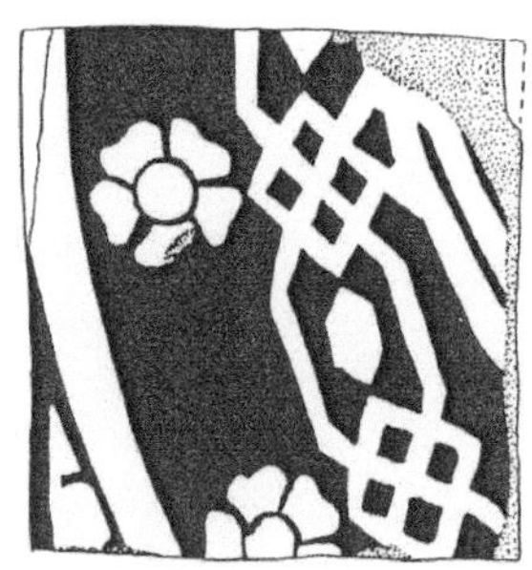

GS21

Leominster Tiles

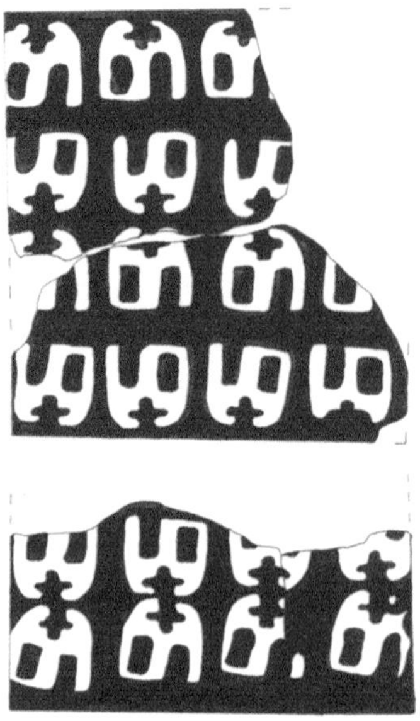

BW 20-22 appear to be unique to Leominster. Judy Stephenson of Hereford Museum suggests these were produced by impressing the bit end of a medieval padlock key as seen in the examples. It is clear that these imprints were made individually from the different juxtapositions on the tiles found.

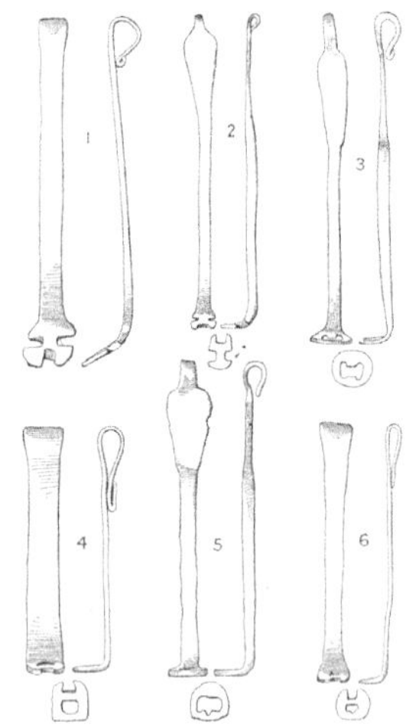

Examples of medieval padlock keys (courtesy Judy Stephenson).

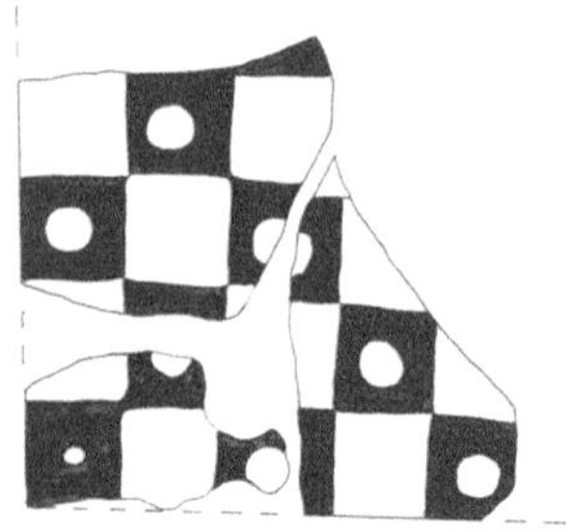

BW 23-25 would again appear to be unique to Leominster, having dots (or counters) within the black squares of a chequerboard design. Laurence Keen confirms he has not seen this elsewhere (2007, pers com).

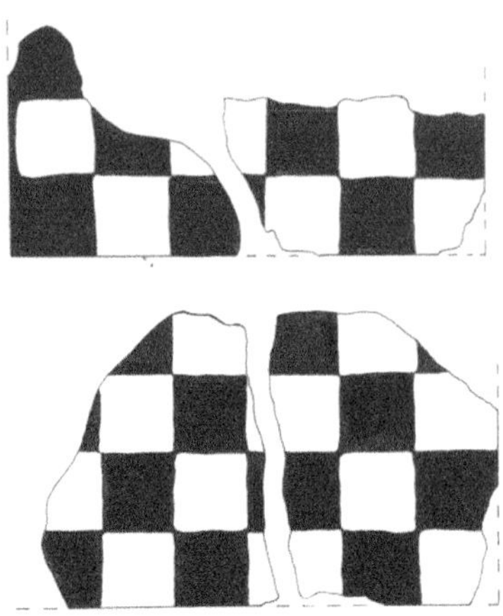

BW 26-28 bear the more common plain chequerboard designs.

Similar tile (GS24) found by Gilbert Scott.

Also at Croft, St Michael's and Brampton Bryan, and at Dilwyn there is a tiny fragment of a chequerboard tile with very much smaller squares.

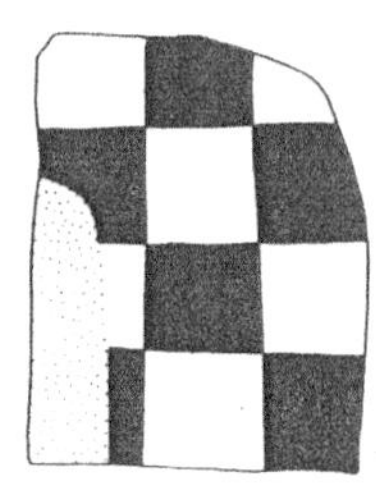

GS24

Leominster Tiles

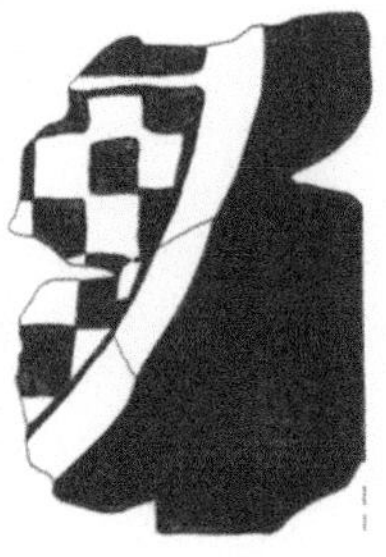 BW 29	**BW 29** consists of three pieces of tile which clearly form part of an heraldic shield. Similar tiles were found by Gilbert Scott (GS24) and by Border Archaeology in 2002.	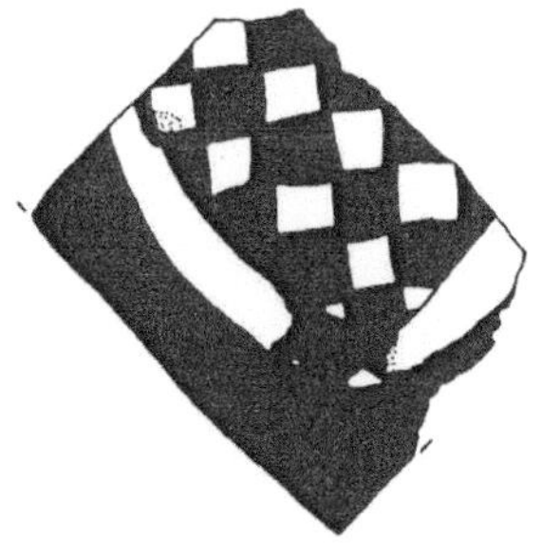 GS24

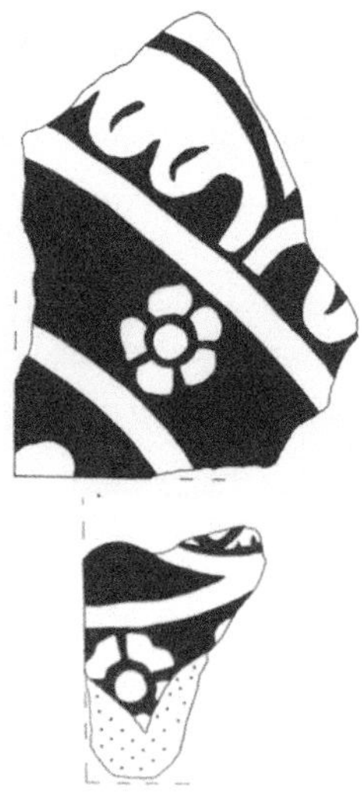 BW35	This design appears to match the description of a tile (fragment 7) found by Border Archaeology in 2002. Also at Croft, St Michael's and Brampton Bryan.	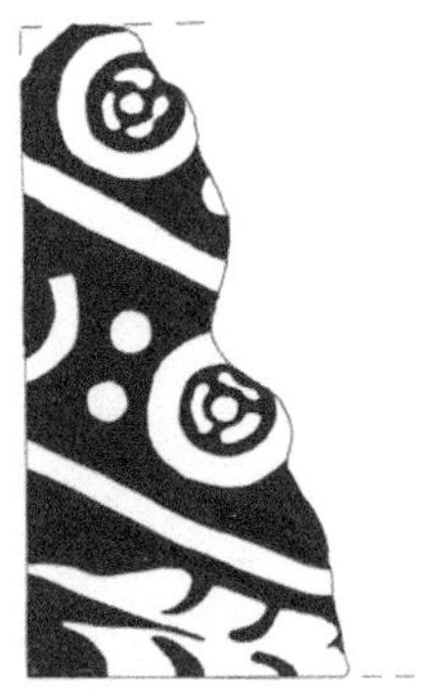 BW36	Also at Croft, St Michael's and Brampton Bryan.

Unidentified fragments:

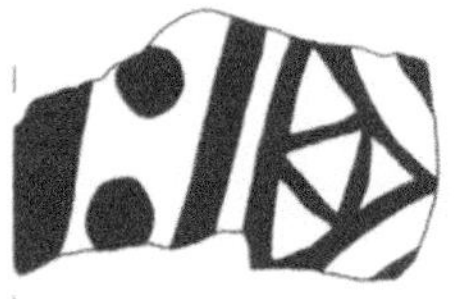 BW31	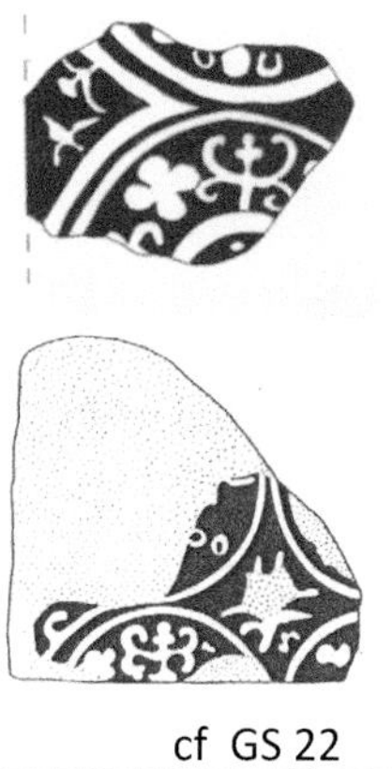 BW32 cf GS 22	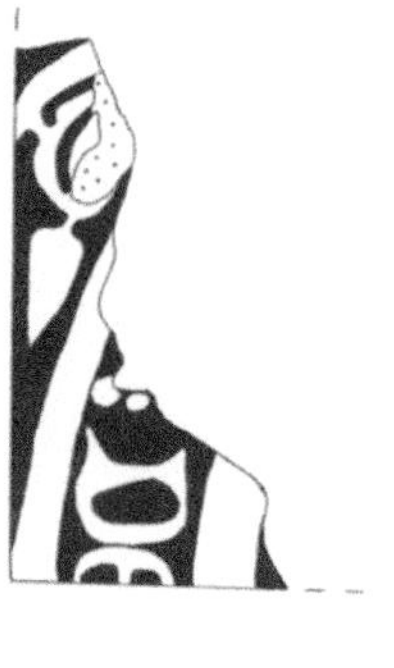 BW33	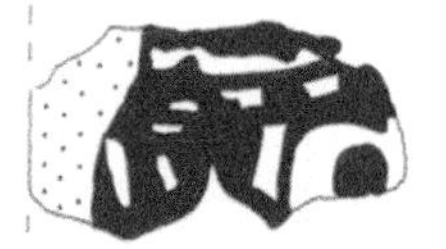 BW34

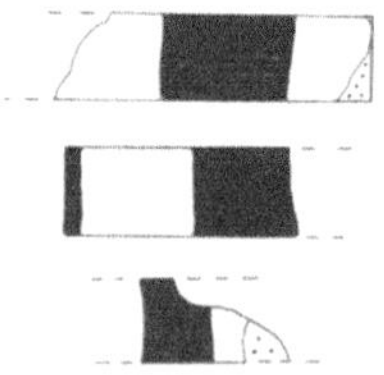 BW50-52	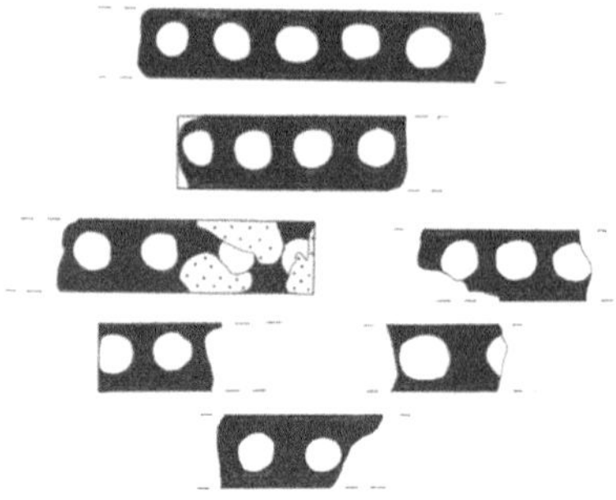 BW53-59	These edging strips are particularly interesting. They measure 15mm wide and have been made individually – not half-scored to be split after firing as would be expected. One fragment of the dotted strip is at Brampton Bryan, and records from Goodrich Castle mention a 'border tile with five or six circles between parallel lines'.

It should be noted that the only tile within the British Museum Compendium attributed to Leominster features a double Tudor rose, but is unlike any other tile known from the site.	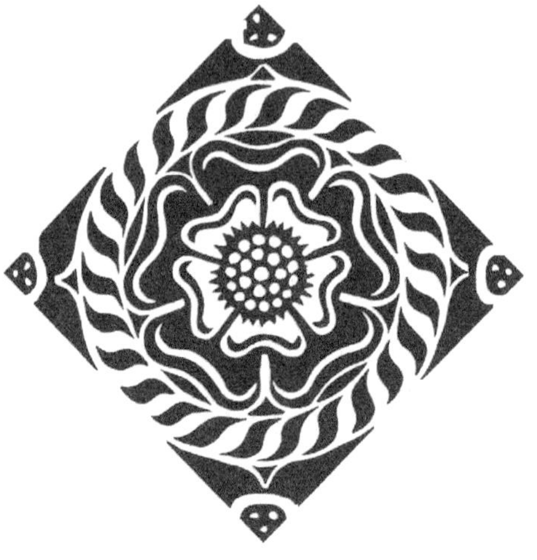 *After Eames:* BM 3096.	An important feature of this single tile is that it is recorded as having five small square stabbed keying holes in its base. The only keying holes found in the 2005 tiles appear to match this description.

Madley - Parish Church

Madley is reputedly the birthplace of St Dubricius, who became a Bishop of Llandaff. The Normans built the first stone church here circa 1100; this was enlarged in 1250 and again in 1320. The monks at Madley were linked to their 'mother church' at Hereford. There was a statue of the Virgin Mary which was believed to have special powers which attracted pilgrims, and this statue was located in the crypt.

In Jane A Wight's book 'Medieval Floor Tiles' 1975 (p.155) she refers to ' ... an owl remaining, with an otherwise rather pathetic collection set in the steps down to the crypt, at Madley Church in Herefordshire'.

Unfortunately the medieval tiles on these crypt steps are now so worn it is virtually impossible to make out any design, let alone an owl.

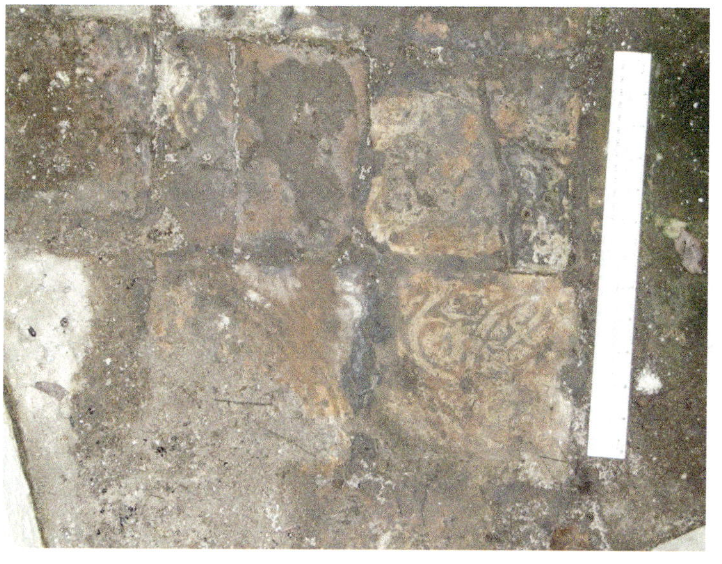

I believe it to be this one from Nottingham in the British Museum Catalogue 2766, not identified elsewhere in Herefordshire.

	*After Eames: BM 2766* Ulverscroft Priory (Nottingham) Chilvers Coton . 14th century.

Monkland - Church of All Saints

Dating from circa 1100, this church was built as a Benedictine cell of the Abbey of Conches in Normandy. It was restored in the nineteenth century. In the nave, on two sides of the pulpit, are two strips of tiles. On the south side are two double-headed eagles within a circle; the only parallels found in Herefordshire to date are in the Booth Chapel of Hereford Cathedral. Flanking these two tiles to fill the required space are segments (one-third) of the tile depicting the arms of Beauchamp.

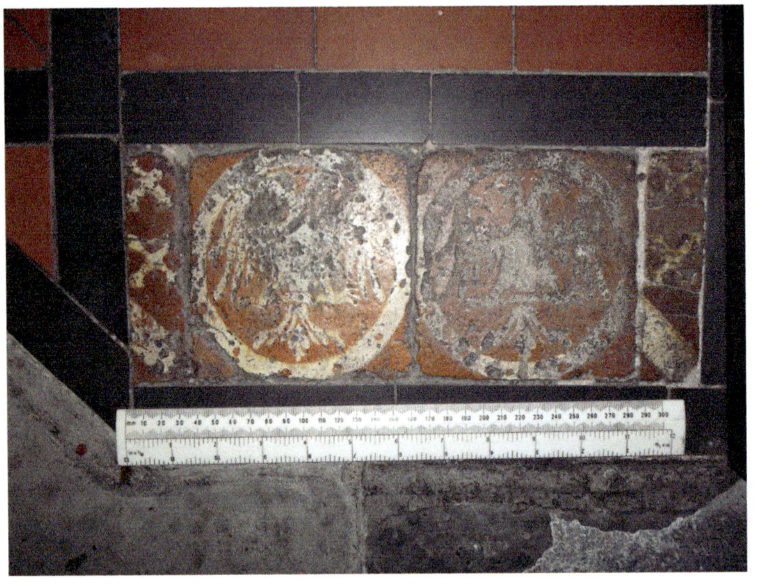

On the west side are five tiles with again a one-third strip of a Beauchamp tile to fill the allotted space. These five comprise three of the square Beauchamp arms; one double-headed eagle within a circle; and one with a white foliate design with a central six-lobed flower.

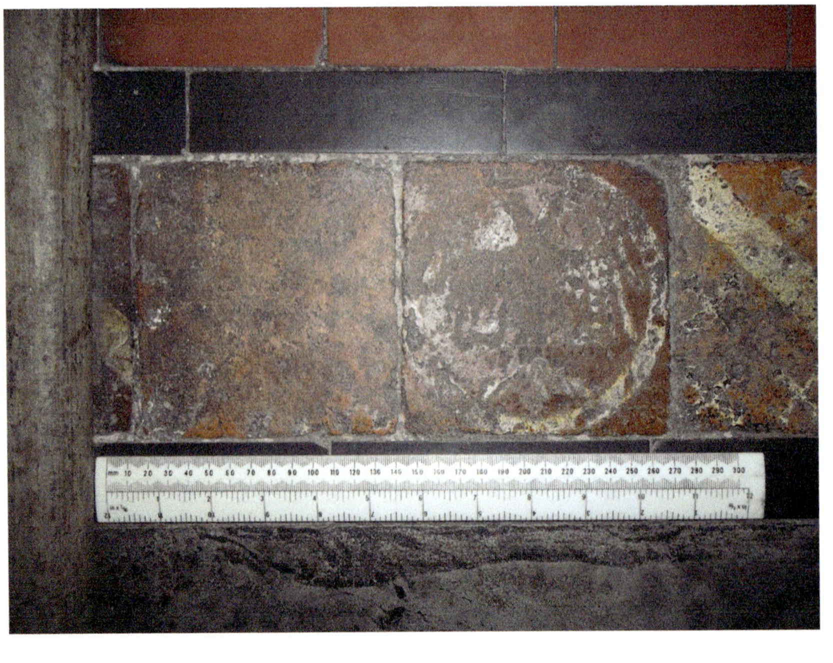

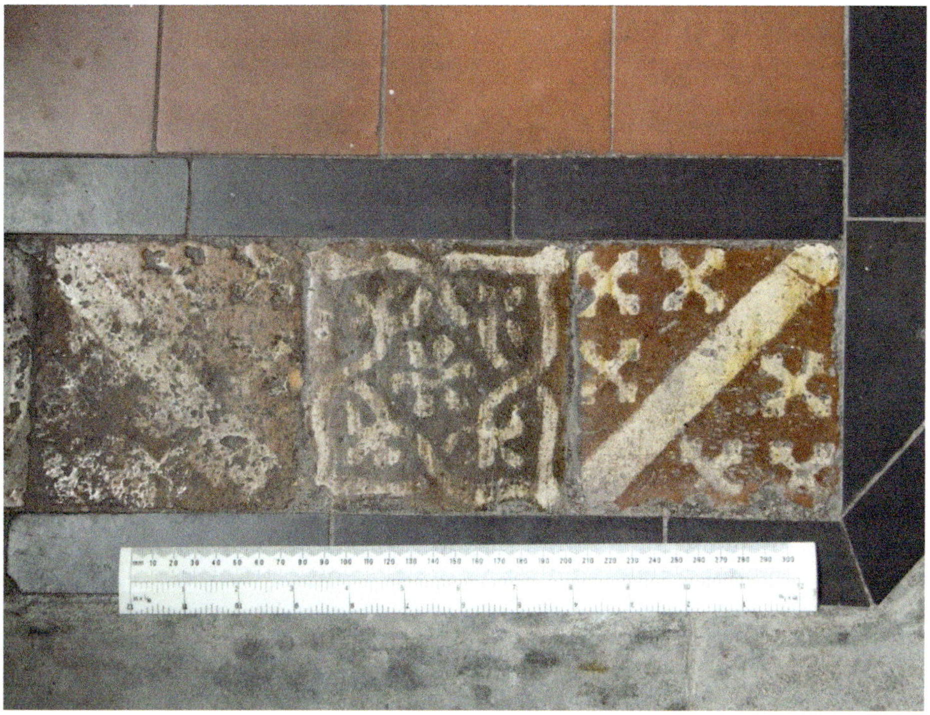

	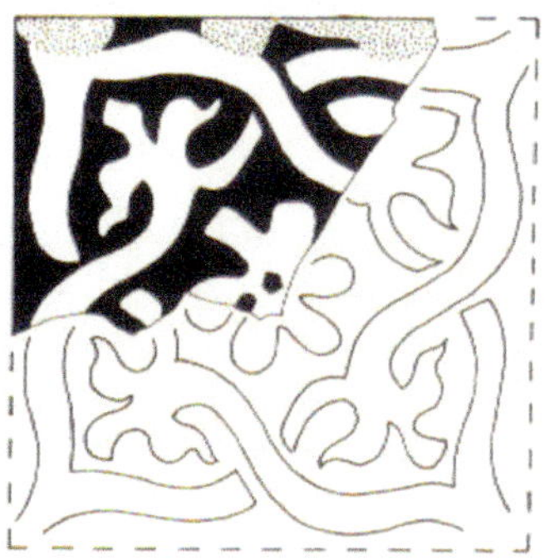 267	
Arms of Beauchamp. Also at Blackfriars, Gloucester, and Worcester Cathedral.	JM Lewis 267 Group 20 BM 2385 Also at Worcester Cathedral.	Double-headed eagle from Booth Chapel, Hereford Cathedral (slightly different feet).

Much Marcle – Hall Court

Alan Hunt carried out an excavation at Hall Court, during which fragments of decorated and plain floortiles were found in two areas; on the lip of the moat to the south of the kitchen, and around the northern structure. The quality would suggest they came from a religious setting, but there is no obvious origin nearby.

He describes three Bredon type straight-set heraldic shield tiles. I have used the drawings from Vince and Wilmott's paper on 'A lost tile pavement at Tewkesbury Abbey', together with Tony Wilmott's descriptions.

 22	*Or, five chevrons azure.* The arms of the Dean and Chapter of Hereford Cathedral. The Clare family crest has three chevrons.
 36a	*Azure, a cross patonce between five martlets, or.* The arms of Edward the Confessor. (Also used by Westminster Abbey, the mother house of Gloucester Abbey, and Malvern.)
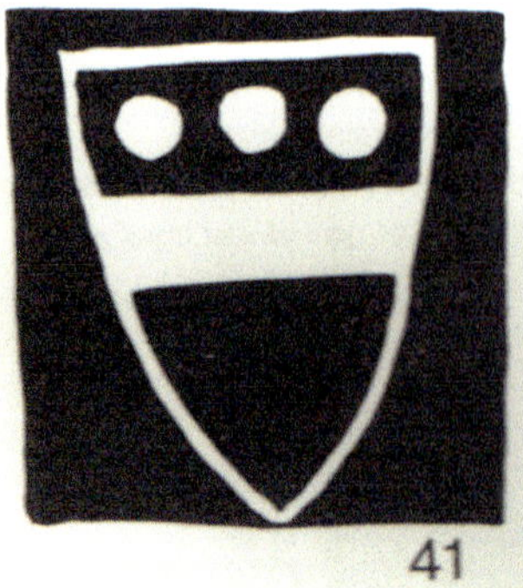 41	*A fess, in chief three roundels.* Possibly William de Devereux. The one found at Much Marcle has three large dots in the lower half of the shield.

The following three designs are also found at Blackfriars, Gloucester; two are also in Hereford Cathedral's Booth Chapel. These are not drawings of the actual tiles found at Hall Court, but those from Blackfriars, Gloucester, drawn by Laurence Keen.

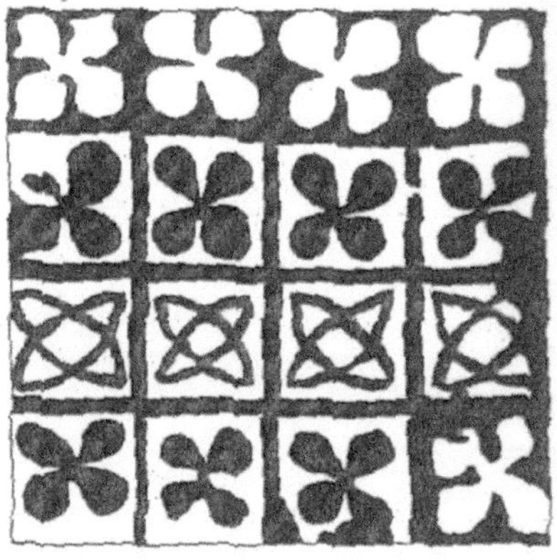	Also at: Blackfriars, Gloucester Dilwyn Hereford Cathedral Booth Chapel Tewkesbury Bredon.
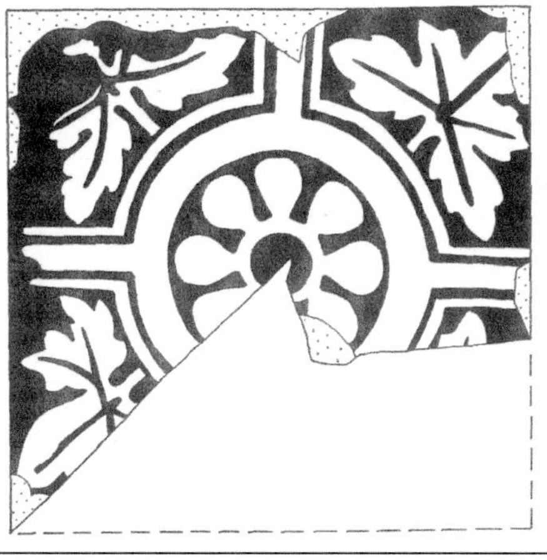	Also at: Blackfriars, Gloucester Hereford Cathedral Booth Chapel Colwall (half-tile).
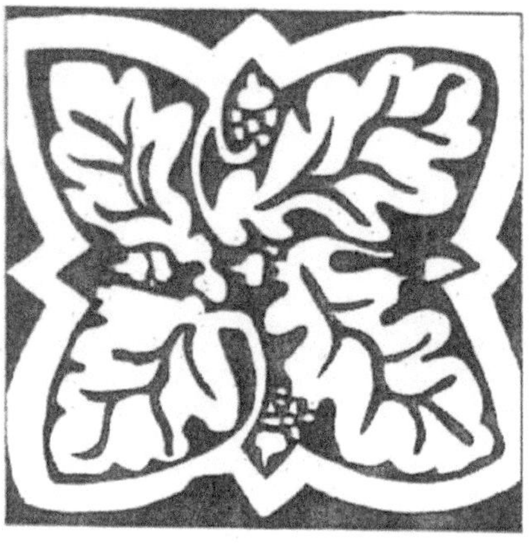	Also at Blackfriars, Gloucester. This tile from Breinton is in the collections of the Hereford Museum. 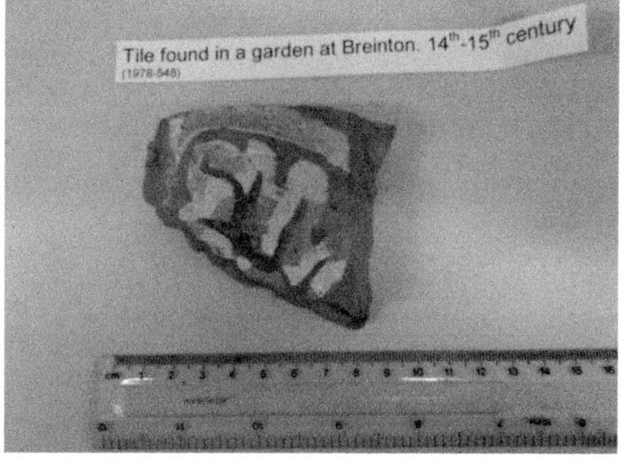

Hall Court tiles - drawings by Alan Hunt, the excavator

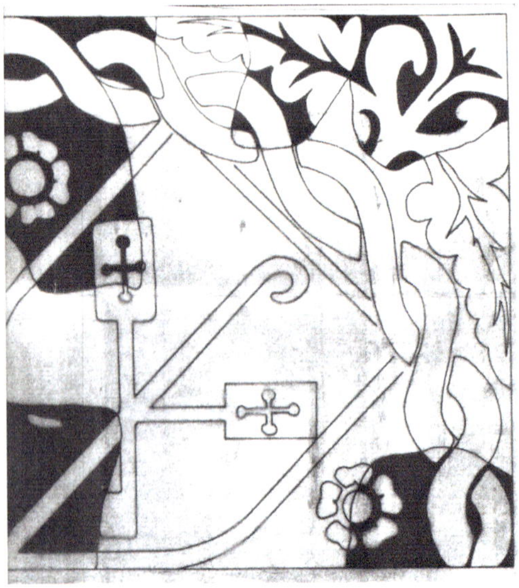

The tile from Much Marcle has the same cable work as this Brampton Bryan tile (below) but the shield is that of Gloucester Abbey (Westminster?) featuring a sword through the two keys of SS Peter and Paul. The corner leaves are not mottled and there is a five lobed flower at either side of the shield.

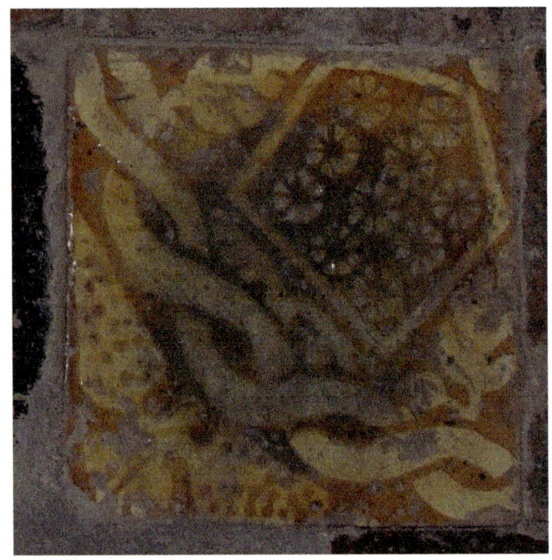

Brampton Bryan.

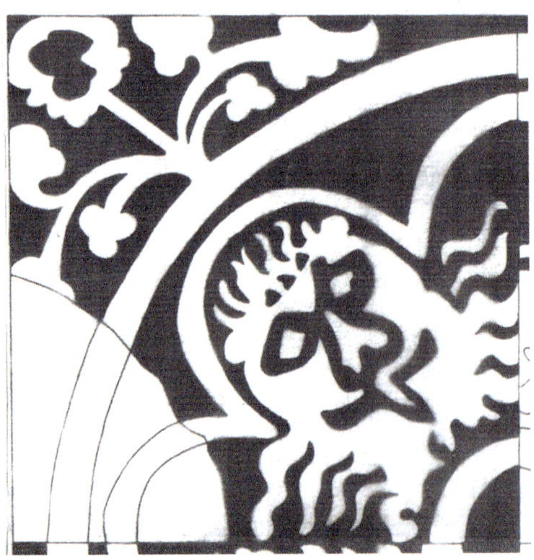

A 4-tile design depicting a lion's head with flowing beard/mane. Alan Hunt comments a similar design is known from Greyfriars, Gloucester.

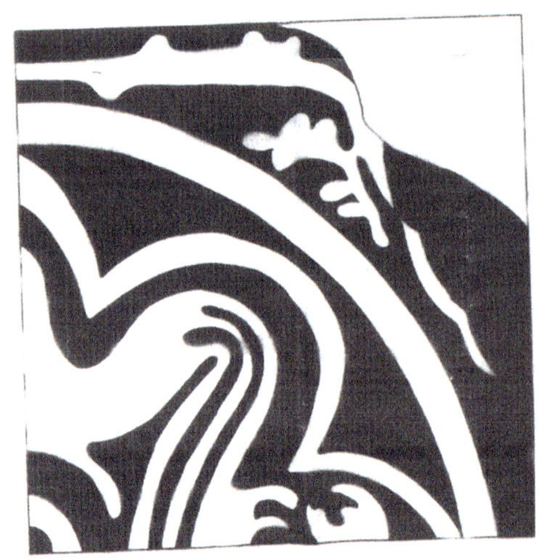

Unidentified.

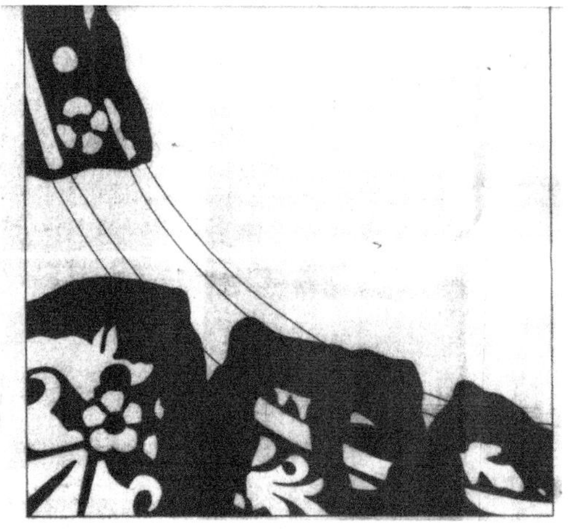

Unidentified.

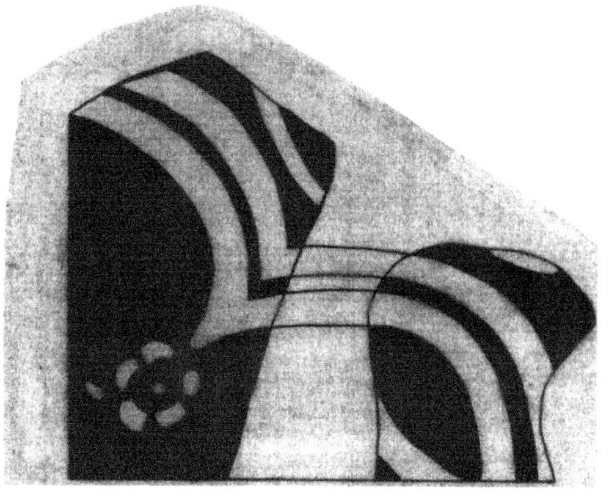

Unidentified.

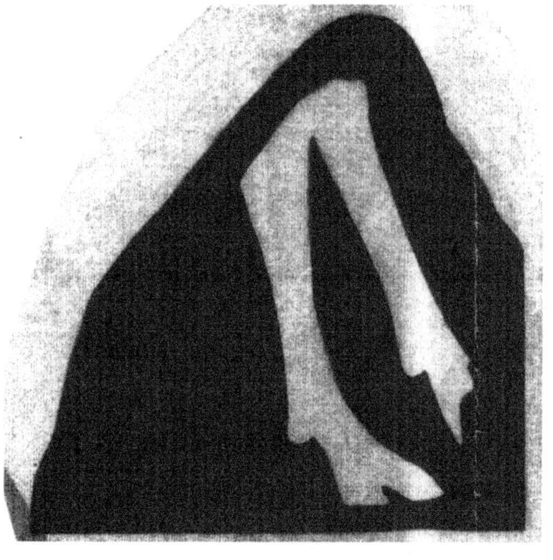

Unidentified.

Stretton Sugwas – The Parish Church of St Mary Magdalene

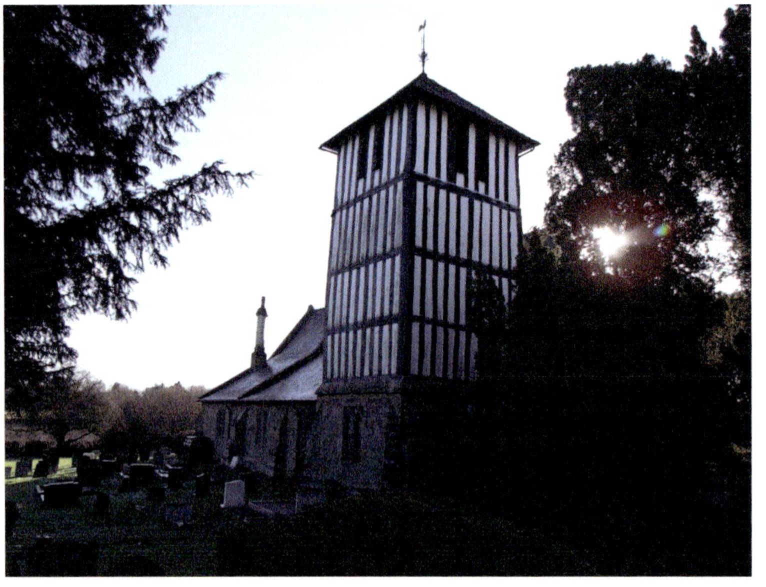

The original church in Stretton Sugwas was built in 1150 in the grounds of what is now the New Priory Hotel. This was in such a state of disrepair that in 1877 the decision was taken to build a new church on a different site but re-using as much as possible of the original materials. Some of the medieval floortiles have been relaid in the vestry.

The similarities with the tiles from St Michael's, Croft, are overwhelming. Alan Vince believed that the tiles at Croft, Stretton Sugwas, and All Saints at Hereford were supplied from Great Malvern or a daughter tilery between 1450-1460. One tile (324) bears the date of 1456.

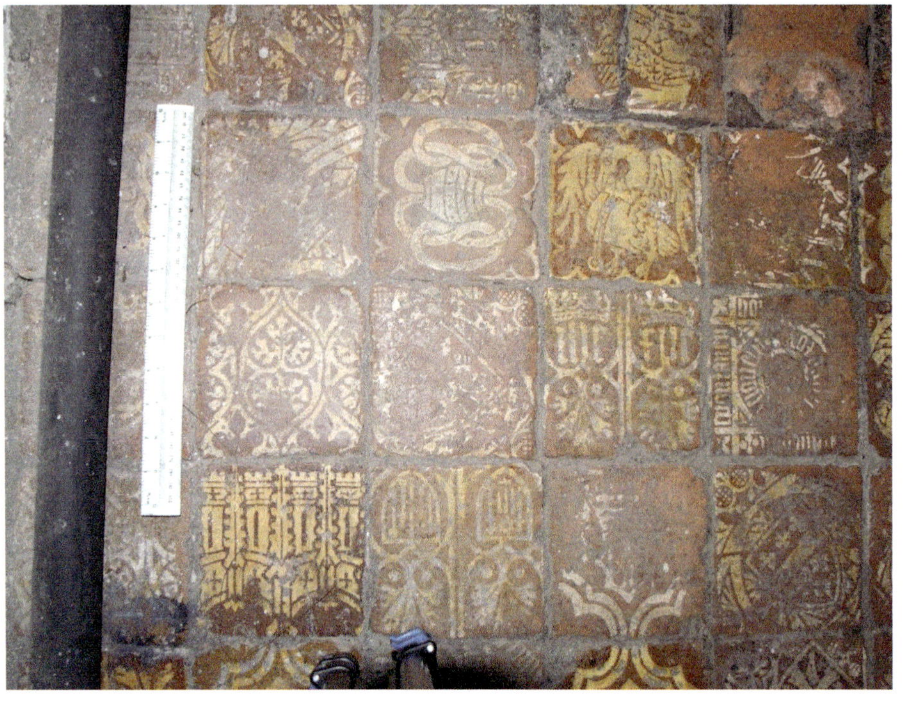

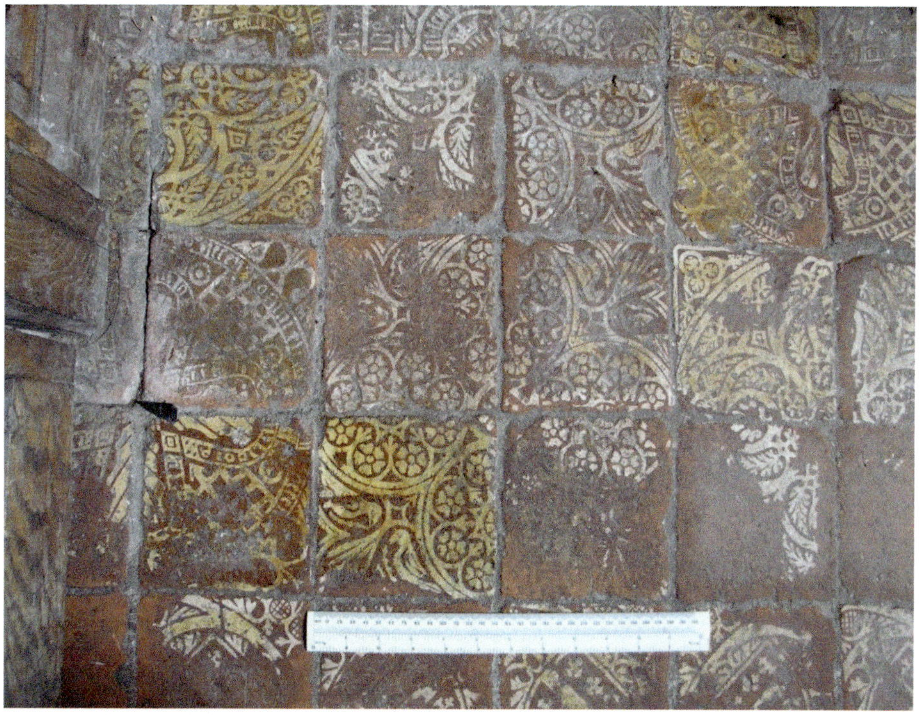

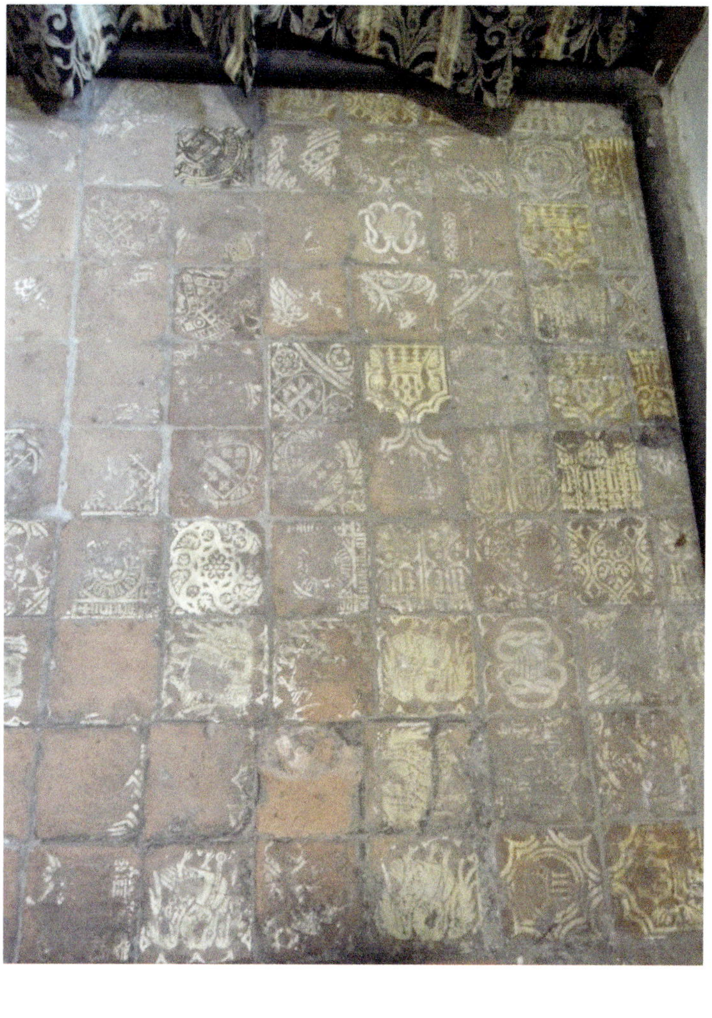

Using the JM Lewis tile numbers, the following tiles are to be found both at St Michael's, Croft, and at Stretton Sugwas:

304, 307, 308 (with a different centre) 310, 312, 318, 322, 324, 332, 333, 337, 338, and 342.

Stretton Sugwas alone has the following tiles (not at Croft):

311 (the crowned M),
320 (the Beauchamp arms)
331 (turrets and pinnacles)
334 (with instruments of the passion)
339 (design with circles and two birds at the corner)
341 (intricate tracery design)
343 (the chained swan of ?The Mohuns ? or the Stafford family)
345 (portcullis with text in circular band)

Croft alone has 315, 317, and 323 (not at Stretton Sugwas).

I have used the drawings and descriptions from JM Lewis' book to identify the designs.

Group 21
304-321 - related to a pavement commissioned for Gloucester Abbey by Abbot Sebroke c. 1455.

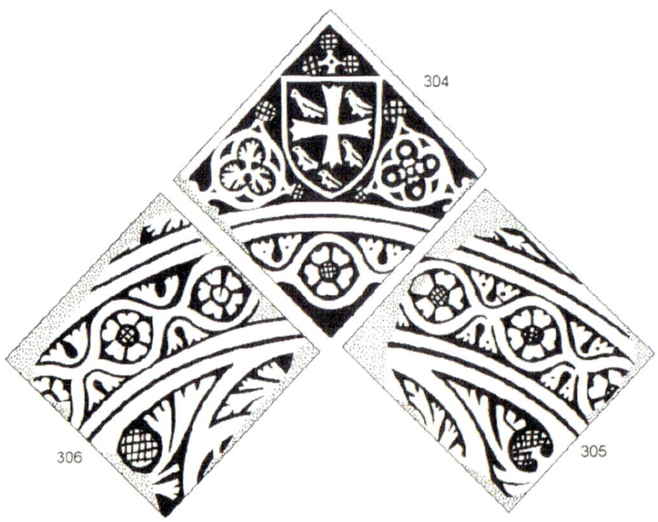	304-6 Slip on impression: 16-tile design: circle of linked five-petalled roses flanked by double lines enclosing a quatrefoil containing leaf designs; outside the circle in each corner a shield of arms; *a cross patonce between five martlets* (Westminster Abbey, the mother house of Gloucester Abbey), flanked by two traceried roundels containing a trefoil and a quatrefoil respectively. 304 = BM Design 2897 305 = BM Design 2898 306 = BM Design 2899 (Great Malvern Priory) Herefordshire: St Michael's, Croft Also at: Raglan Castle, Monmouth.

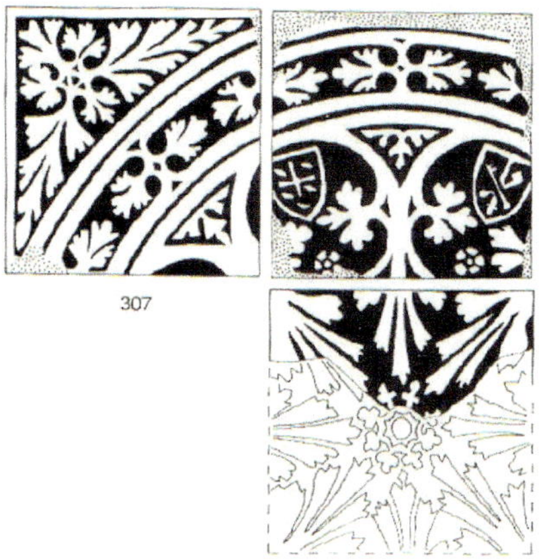

307 308 309

336

307-9

Slip on impression:
9-tile design; circle of double lines flanking a frieze of opposed three-leaf motifs: within the circle leaf tracery forming eight compartments containing shields of arms of two designs arranged alternately: *a cross patonce between five martlets* (Westminster Abbey) and *cross keys with sword in pale* (Gloucester Abbey); in the centre an octagonal motif of linked trefoils; outside the circle each corner has a framed leaf motif.

307 = BM Design 2859 (Great Malvern)
308 = BM Design 2860 (Gloucester Cathedral, Great Malvern)
309 = BM Design 2565 (Great Malvern)

Herefordshire: St Michael's, Croft
Stretton Sugwas has different central tile
336 = BM 2435.

Also at: Monmouth, Abergavenny , Llangattock-nigh-Usk.

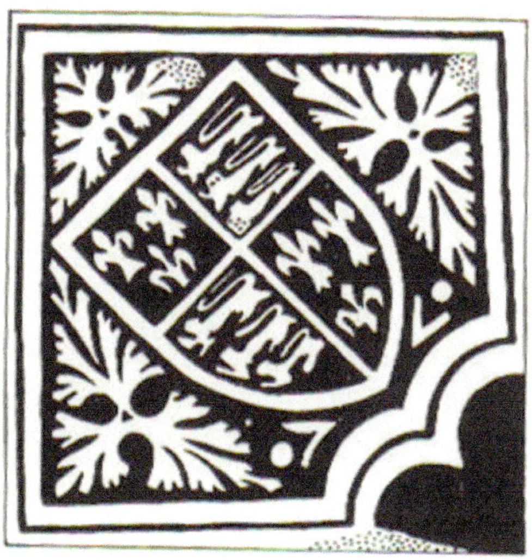

310

310-15

Slip on impression:
4-tile or larger arrangement consisting of a grille of reserved quatrefoils with the arms of England at the corners; the designs between the quatrefoils in the centre vary (312-15).

310 Slip on impression:
Shield of arms set diagonally : *quarterly 1st & 4th three fleur-de-lys 2nd & 3rd three lions passant guardant* (royal arms of England, post 1405) within plain double border with quarter quatrefoil in bottom corner; the corners within the border are filled with foliage decoration.

310 = BM Design 1518
(Gloucester Cathedral, Great Malvern Priory, Lenton kiln site).

Herefordshire: St Michael's, Croft
Also at: Abergavenny, Monmouth, Raglan.

311B

311 Slip on impression:
9-tile or larger arrangement; crowned M within double border, for Henry VI's queen, Margaret of Anjou (d.1482); flanked by foliage motifs; two quarter-quatrefoils below.

311 = BM 1421
(Gloucester Cathedral, Great Malvern Priory)

Herefordshire: St Michael's, Croft.

Also at: Raglan, Monmouth.

312

312 Slip on impression:
Centre tile; crowned monogram *ihc* within double circle; quarter quatrefoils in each corner.

312 = BM Design 1418
(Great Malvern Priory)

Herefordshire: St Michael's, Croft.
Also at: Monmouth.

318A

318 Slip on impression
4-tile design; a barbed quatrefoil frame with repeating inscription: *fiat voluntas dei* (May the Will of God be done), surrounding four shields of arms; *three lions passant guardant* (royal arms of England, pre-1340); outside the frame each corner has a pair of birds flanking a central stem.

318 = BM Design 1480
(Gloucester Cathedral, Great Malvern Priory)

Herefordshire: St Michael's, Croft
Also at: Carmarthen, Raglan, Llangattock-nigh-Usk, Monmouth

320

320 Slip on impression:
4-tile design; within a circle four shields of arms set diagonally each within a ring of stylized foliage: a fess between six cross crosslets (Beauchamp); outside the circle each corner has a stylized foliage motif.

320 = BM Design 1601
(Great Malvern Priory, Lenton Priory kiln site)
Also at: Llangattock-nigh-Usk, Llantilio Pertholey, Monmouth, Raglan.

322

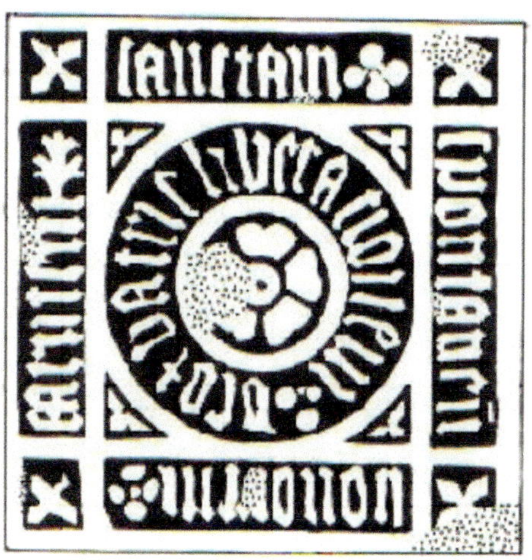

322 Slip on impression:
Square subdivided into a central square by narrow rectangular borders; the centre has a central five petalled rose surrounded by a circular band; the borders and central ring are filled by a continuous black-letter inscription reading: *mentem sanctam, spontaneu(m) honorem deo, et patrie liberacionem* (A holy mind, honour freely given to God, liberty to the country).
These words refer to St Agatha and occur in her *Life* (Bollandius & Herschinjus 1863, 640A). They were used in the Middle Ages as a charm against fire.

322 = BM Design 1429
(Great Malvern Priory, Lenton Priory kiln site)
Herefordshire: St Michael's, Croft
Also at: Monmouth, Raglan.

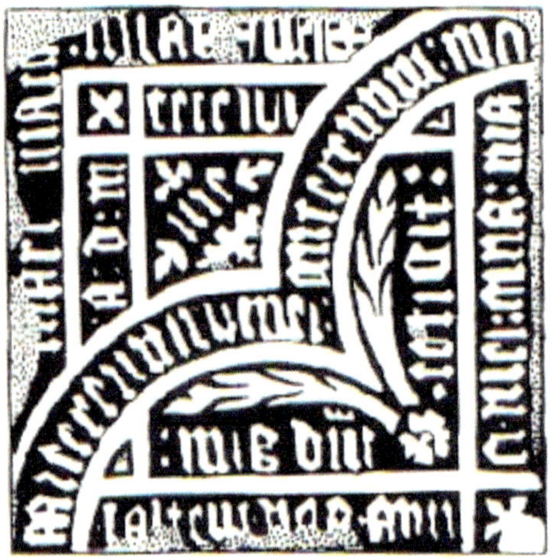

324

324 Slip on impression:
4-tile design; ribbon quatrefoil with double border outside and equal-arm cross inside; all the spaces are filled with a continuous black-letter inscription repeated four times: *marc mathe:lucas:joh/ a:d:mccccIvi/ miseremini:mei:miseremini:mei/ saltem.vos.amici.mei:quia:ma/nus dni.tetigit:/me* (Mark, Matthew, Luke, John. A>D> 1456. Have pity upon me, have pity upon me at least ye my friends, for the hand of the Lord hath touched me).
The words include a passage from Job xix,21 used in the Office of the Dead.

324 = BM Design 1468
(Great Malvern Priory, Baynham Abbey)

Herefordshire: St Michael's, Croft
Also at: Llangattock-nigh-Usk, Llanthony Priory, Monmouth, Raglan.

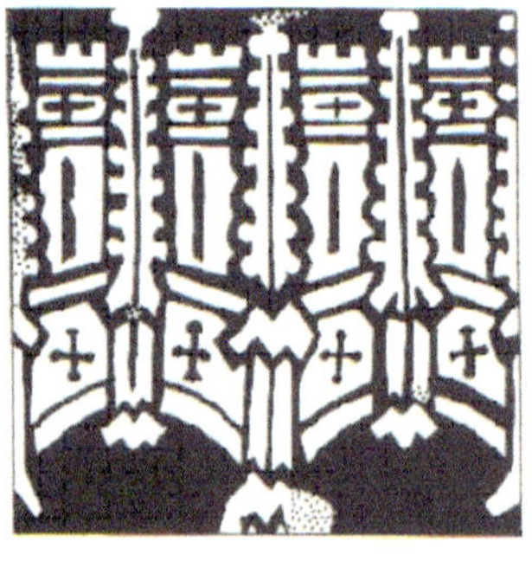

331

331-4 Square wall-tile series based on tabernacle-work, made originally for Great Malvern Priory.

331 Slip on impression:
Turrets and pinnacles.

331 = BM Design 2577
(Great Malvern Priory)

Herefordshire: Hereford, All Saints
Also at: Monmouth, Raglan.

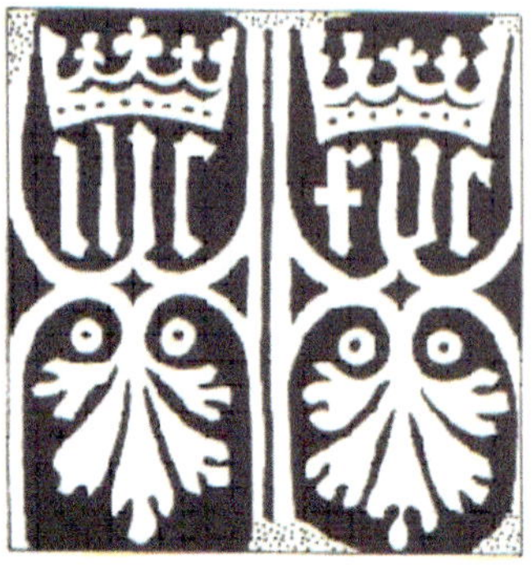

332

332 Slip on impression:
Two-light opening with pendant leaf motifs, and with crowned monograms above: *ihc* and *xpc* (Jesus/Christ).

332 = BM Design 1420
(Great Malvern Priory, Lenton Priory kiln site)

Herefordshire: St Michael's, Croft; Hereford, All Saints

Also at: Abergavenny, Monmouth.

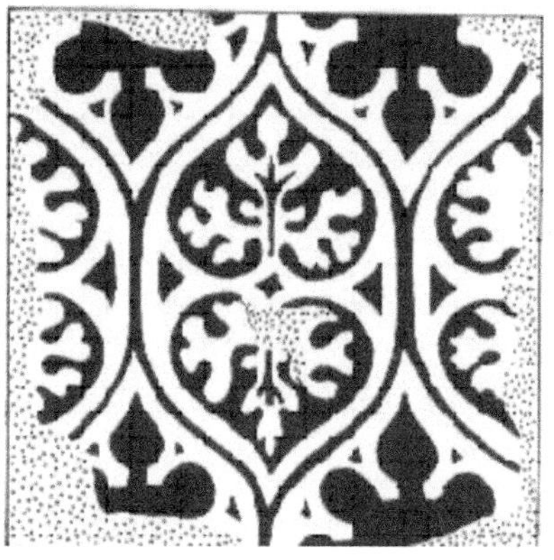

333

333 Slip on impression:
Two-light tracery.

333 = BM Design 2575
(Great Malvern Priory, Lenton Priory kiln site)

Herefordshire: St Michael's, Croft

Also at: Monmouth.

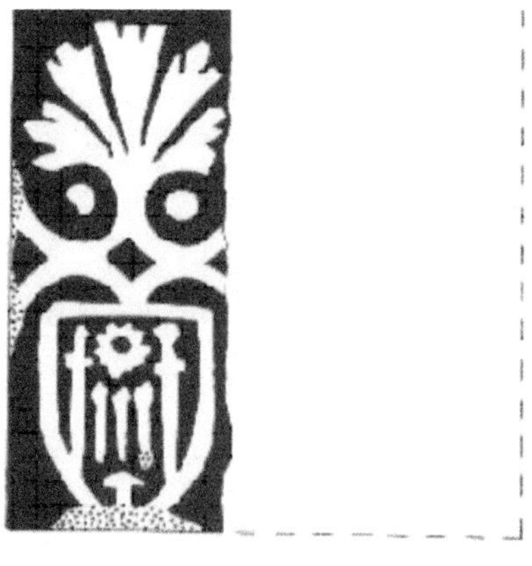

334

334 slip on impression:
Half of a two-light opening, with leaf motif above and shield of arms with some of the Instruments of the Passion below.

334 = BM Design 1718
(Great Malvern Priory, Great Lenton Priory kiln site)

Also at: Monmouth.

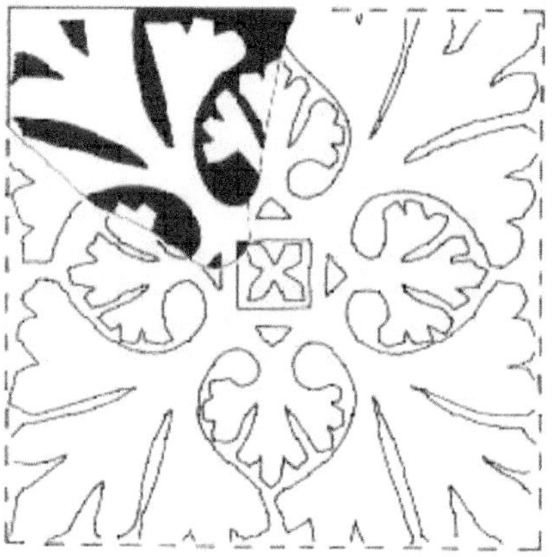

337

337 Slip on impression:
Foliage decoration arranged in the form of a cross set diagonally with smaller areas of decoration between the arms.

337 = BM Design 2533
(Great Malvern Priory, Lenton Priory kiln site)

Herefordshire: St Michael's, Croft

Also at: Raglan.

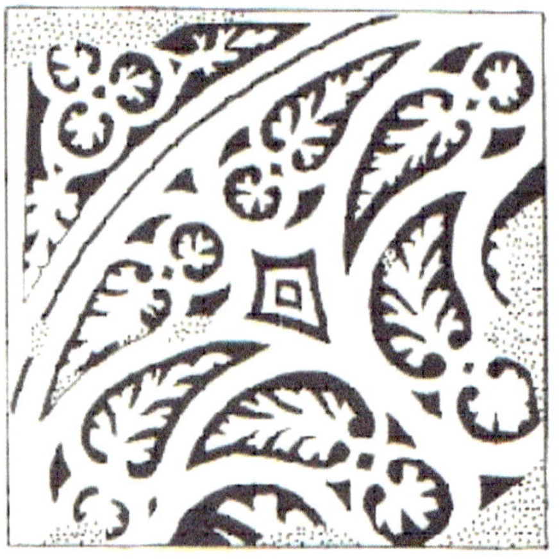

338

338 Slip on impression:
4-tile design; circle with elaborate arrangement of foliage designs within reserved mouchettes; outside the circle each corner is occupied by a trefoil and leaf motif.

338 = BM Design 2689
(Great Malvern Priory, Lenton Priory kiln site)

Herefordshire: St Michael's, Croft

Also at: Monmouth, Raglan.

339

339 Slip on impression:
4-tile design; four circles with inscribed octofoils, each lobe containing a five-petalled rose; in each outer corner two birds flanking a central foliage design.

339 = BM Design 2688
(Great Malvern Priory, Lenton Priory kiln site)

Also at: Monmouth, Raglan.

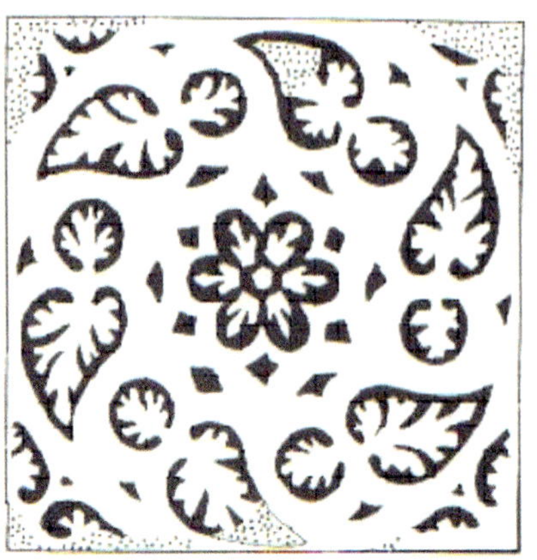

341

341 Slip on impression:
Circle with central hexafoil surrounded by reserved circles alternating with reserved mouchettes, each containing a leaf motif.

341= BM Design 2439
(Great Malvern Priory, Lenton Priory kiln site)

Also at: Llangattock-nigh-Usk, Monmouth.

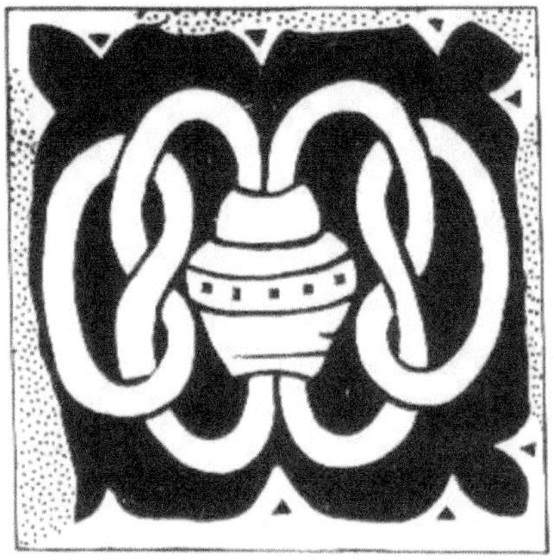

342

342 Slip on impression:
Within a square cusped frame *a Stafford knot surrounding the nave of a wheel.* The badge of the Stafford family, probably for Humphrey Stafford, Duke of Buckingham, d.1483

342 = BM Design 1739
(Great Malvern Priory)

Herefordshire: St Michael's, Croft

Also at: Llangattock-nigh-Usk, Llantilio Pertholey, Monmouth.

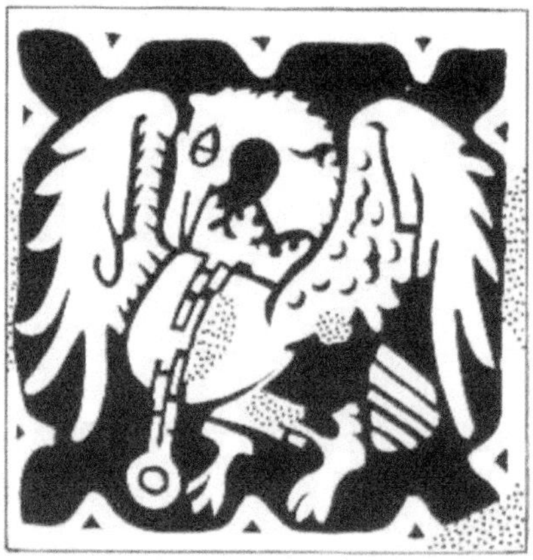

343

343 Slip on impression:
Within a square cusped frame: *a swan with wings overt, collared and chained.* A swan was the badge of the Bohun family which was adopted by Henry of Lancaster (later Henry IV) after his marriage, later becoming a badge of the Prince of Wales (Henry V).
Alternatively, in the view of 342, it might be a badge taken from the crest of the Stafford family (Great Malvern and Little Malvern Priories).

Also at: Monmouth

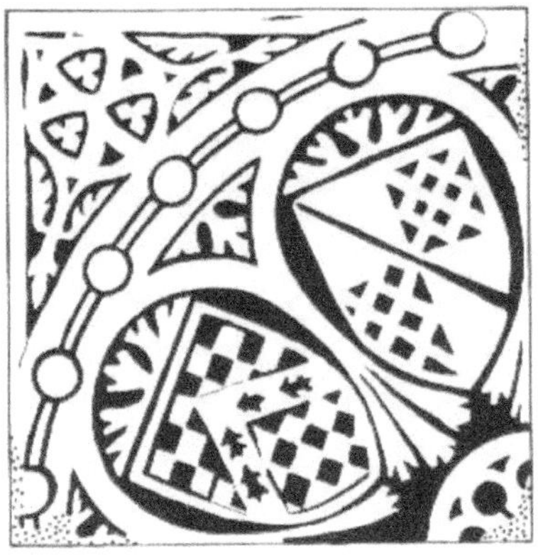

344

344 Slip on impression:
4-tile design: a circle decorated with linked roundels enclosing an octofoil of leaf tracery surrounding a central circle; the lobes of the octofoil contain alternating shields of arms: *chequy a chevron ermine* (Newburgh) and *quarterly 2nd and 3rd a fret over all a bend* (Despenser); outside the circle in each corner a circle divided radially into six, the alternate spaces containing a trefoil.

344 = BM Design 1723 (Great Malvern Priory)

Also at: Monmouth.

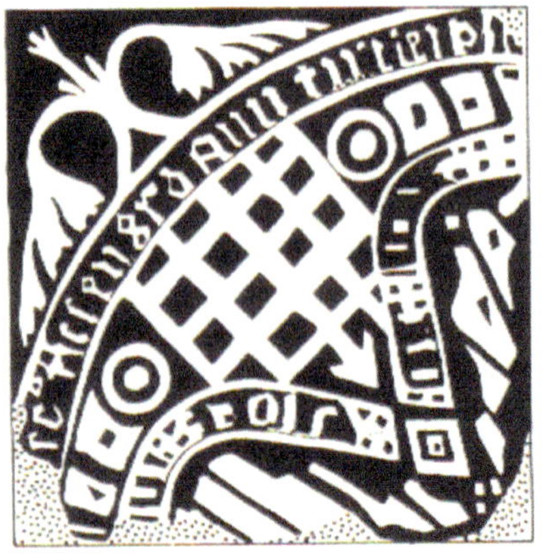 345	345 Slip on impression: 4-tile design; circular band with indecipherable (probably mock) inscription surrounding a quatrefoil outlined by a similarly inscribed band; set diagonally between the circle and angles of the quatrefoil are four portcullis emblems (Somerset); in each corner outside the circle a weeping foliage motif. (Sir Charles Somerset (c.1460-1526), son of Henry Beaufort, 2nd Duke of Somerset, who married Elizabeth, only child of William Herbert, Lord Herbert of Raglan, was the ancestor of the Dukes of Beaufort (Siddons 1993). Also at: Abergavenny, Llangattock-nigh-Usk, Monmouth.

Thornbury – Netherwood

Netherwood was a manor of the Botilers, the de Moignes, passing to the Mortimers in the fourteenth century. The original house was pulled down but some of the earlier material was used in the re-building. The floortile fragments have been curated by an estate worker, Mr Ivan Turner, and are now in a private collection at Netherwood and at Mr Turner's home.

Hilary White of the Hereford SMR viewed the tiles in 1989 and I quote from her report here. '167 fragments of tile were presented and were labelled as being found in the house, the moat, and a number of fields around the estate. No complete tiles survive, but they seem to have had little surface wear. The size of the tiles is about 12cm, with thickness varying from 2.2 – 2.8 cm. The sides are either flat or slightly bevelled and there are no keying holes on the reverse'.

The decoration of the collection closely parallels tiles found at the Worcester Cathedral Singing School.

The drawings of the Netherwood tiles were produced by Hilary White and Laurence Keen. The drawings from Worcester Cathedral were produced by Laurence Keen.

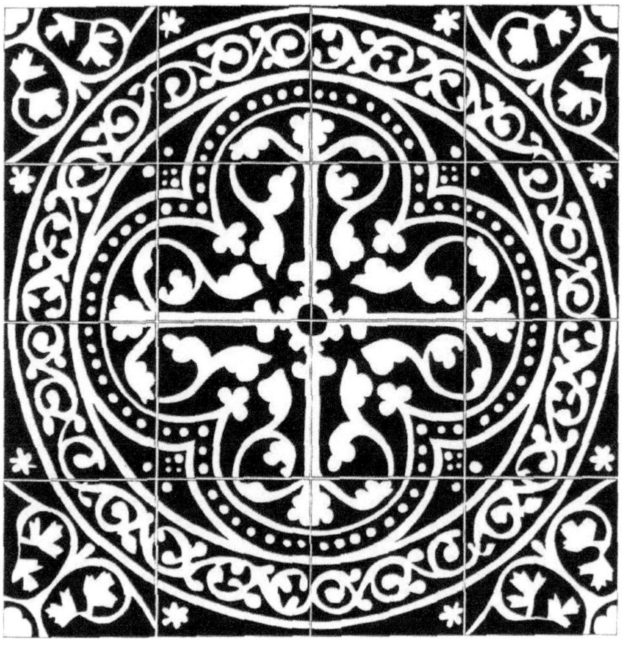	J M Lewis 262-4. 4 tiles repeat to make up a 16-tile design. Two outer circles enclosing stylized foliage with inner double dotted quatrefoil. Stylised foliage in centre and at corners. Also at Worcester Cathedral, Chepstow and Usk.

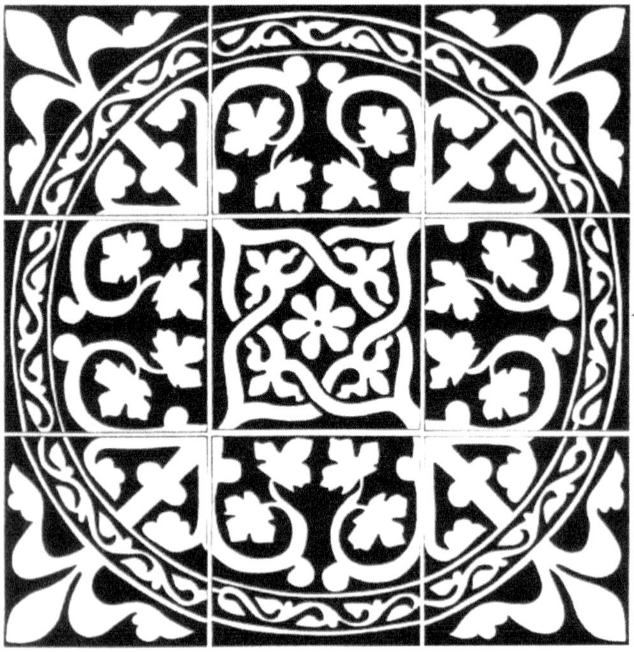

JM Lewis 265-7.

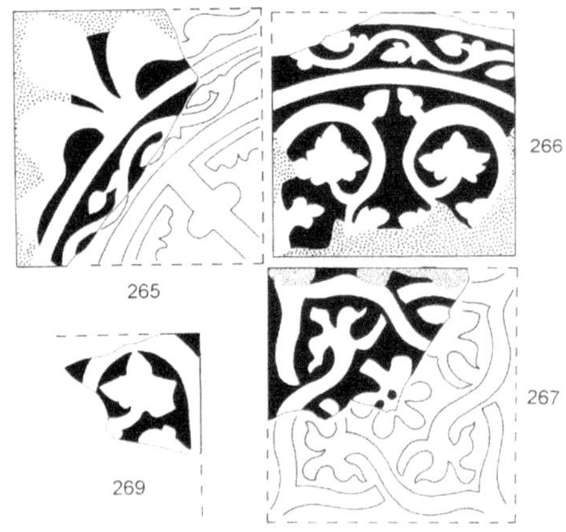

Group 20 (Droitwich Group).

9-tile design; an elaborate cross fleury set diagonally but apparent as a circular band with running leaf design surrounding a foliage design intertwined at the centre; in the corners outside the circle are large fleurs de lys forming the ends of the arms.

Also at Worcester Cathedral, Chepstow Priory, and Blackfriars, Gloucester (but with slightly different corner tiles).

Middle edge 266 at Hereford Cathedral, Booth Chapel but associated with this different corner tile.

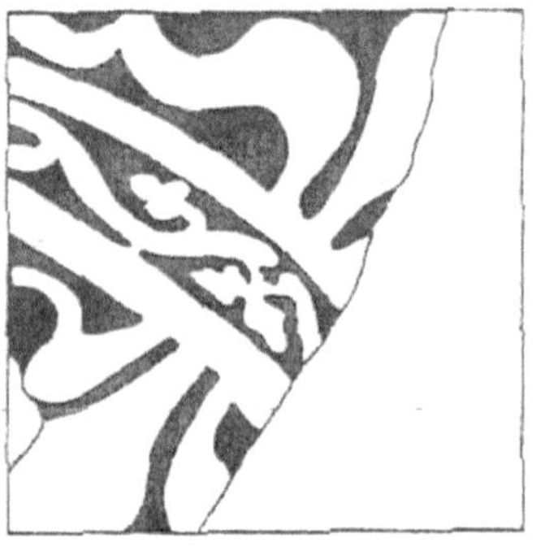

	4-tile design of interlocking circles with stylized foliage. Also at Worcester Cathedral.
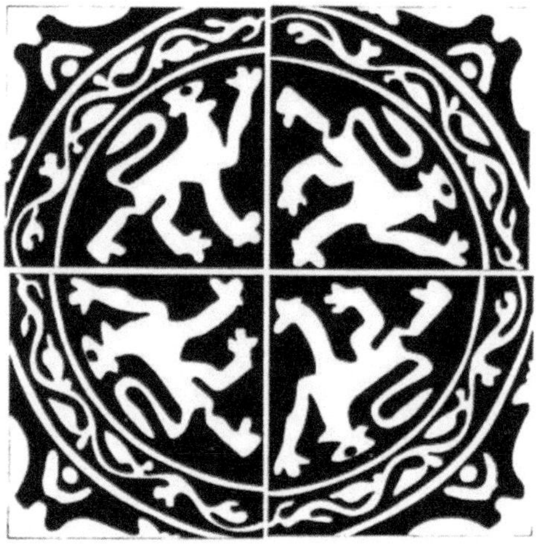	4-tile design; double circle containing flowing foliage with four lions passant within. Also at Worcester Cathedral.
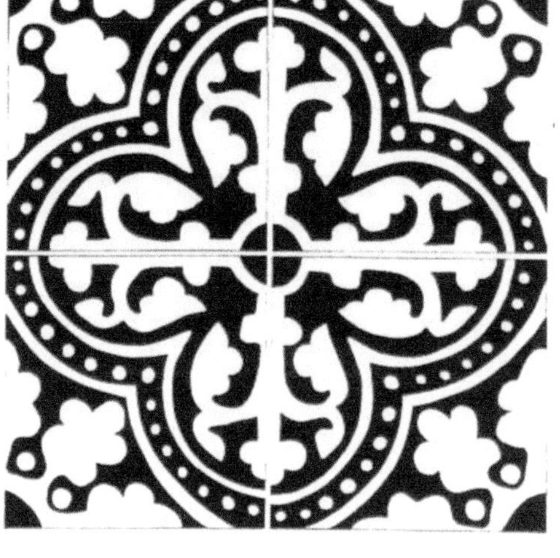	4-tile design; double quatrefoil containing dots enclosing stylized foliage. At each corner a quarter circle of foliage interspersed with dots. Also at Worcester Cathedral.

4-tile design to depict one lion passant; Simplified foliage.

Also at Worcester Cathedral.

Wr 7-10

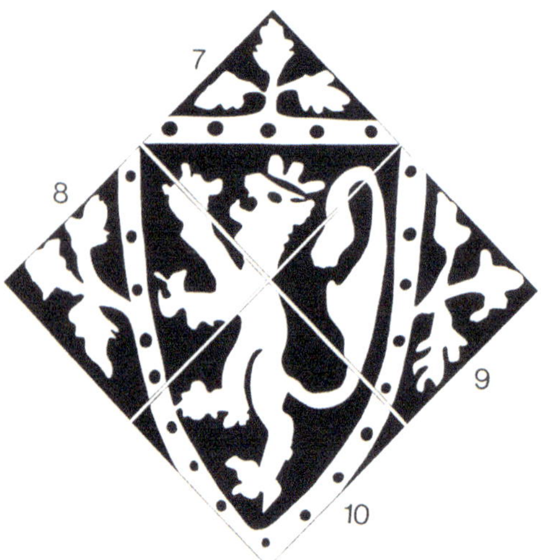

Worcester Cathedral

4 tiles make a crowned Lion Rampant. Dots within border of shield. Arms of Richard of Cornwall?

Also at Worcester Cathedral.

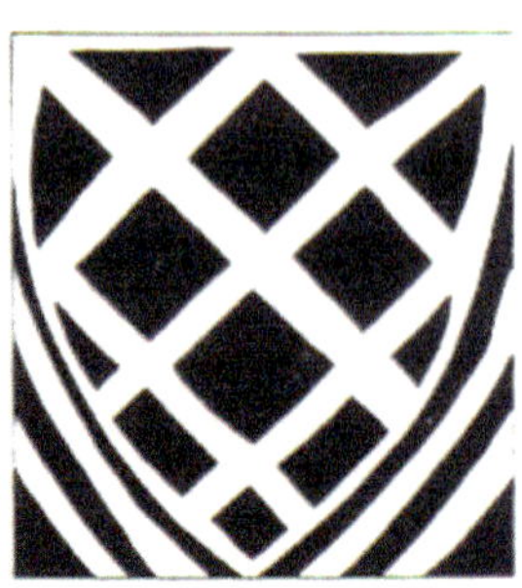

Single tile, upright arms; Fretty (possibly Audley or Verdon)
Also at Worcester Cathedral.

Single tile, upright arms. Stags head.

Also at Worcester Cathedral.

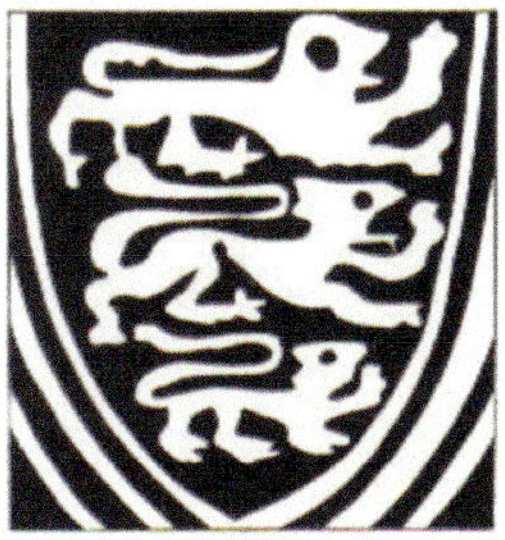

Single tile, upright arms. Triple lion passant sinister. Arms of Old England.

Also at Worcester Cathedral.

Single tile, upright arms. Fleur de lys.

Also at Worcester Cathedral.

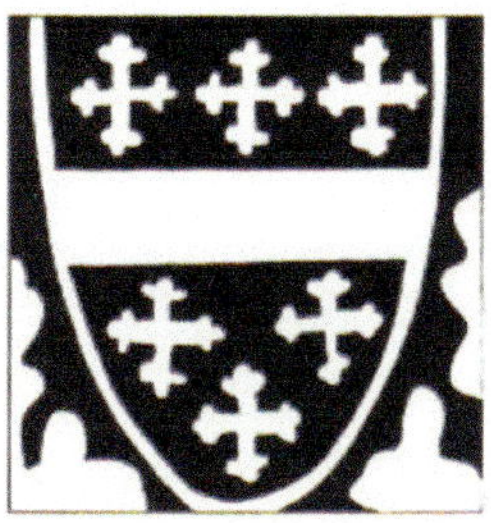

Single tile, upright arms. Fesse, 6 cross crosslets. Beauchamp.

Also at Worcester Cathedral (and possibly Dilwyn and Monkland).

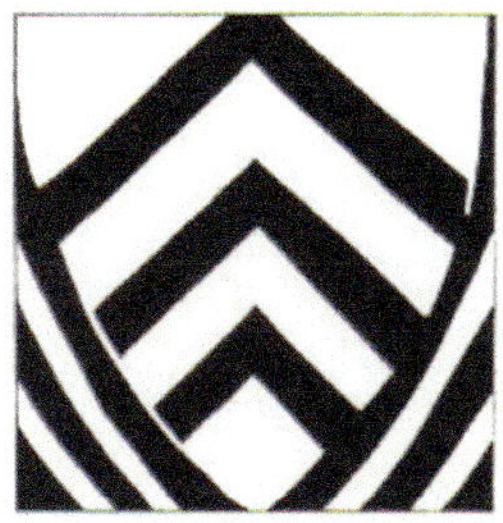

Single tile, upright arms. Triple chevron. De Clare.

Also at Worcester Cathedral.

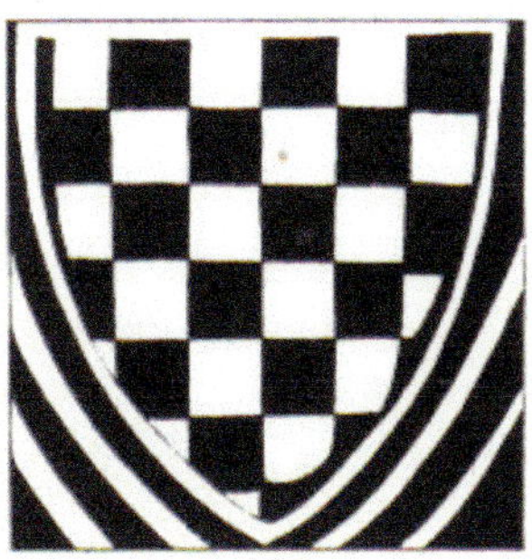

Single tile, upright arms. Checky. De Warenne

Also at Worcester Cathedral.

Hilary White also mentions " a single tile. Upright arms. Fleur de lys. Arms of old France "
Likely to be similar to this example from Hereford Cathedral Booth Chapel:

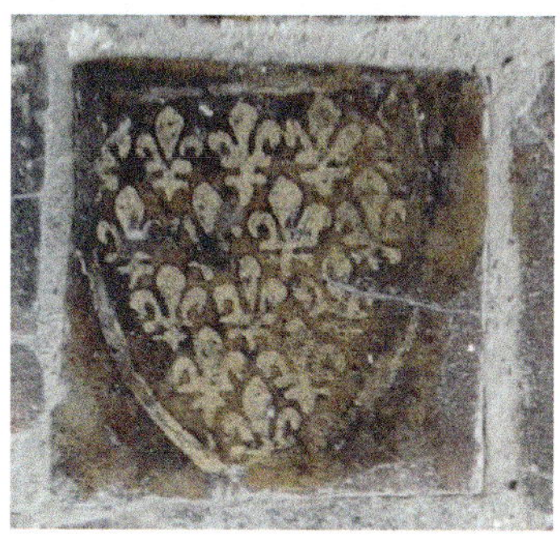

Azure, semee-de-lis or
The arms of France (ancient) adopted by Philip II 1120-80.

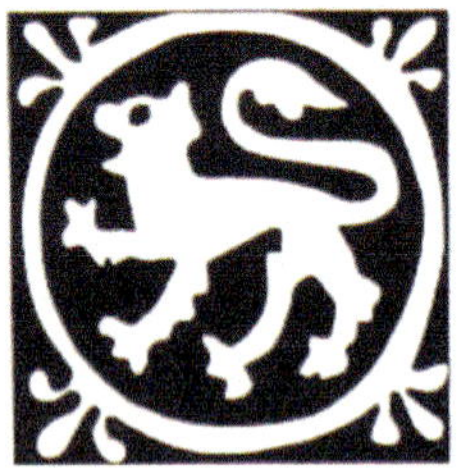	Lion passant within circle; small irregular lys in each corner. The most common surviving tile from Netherwood. Also at Worcester Cathedral, Hereford Cathedral Lady Chapel, Wigmore Abbey, and a crude version from Goodrich Castle.
	Eight petal flower on white square set at 45 degrees to tile. Stylised foliage in corners. Also at Worcester Cathedral, and Hereford Cathedral Lady Chapel.
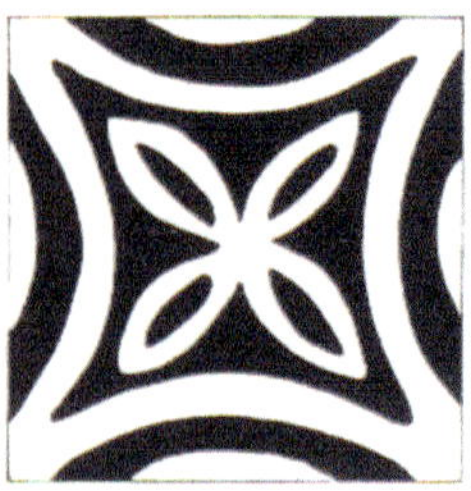	Stylised 4 petal flower at centre. Also at Worcester Cathedral, Hereford Cathedral Lady Chapel, and Wigmore Abbey.
	Decorative fleur de lys set diagonally; extra frond between petals. Also at Worcester Cathedral and Hereford Cathedral Lady Chapel.
 47 Worcester Cathedral	Plain fleur de lys set diagonally.

Westhide – St Bartholomew's Church

There is a single sliptile featuring the arms of SS Peter and Paul reset into the modern chancel screen.

	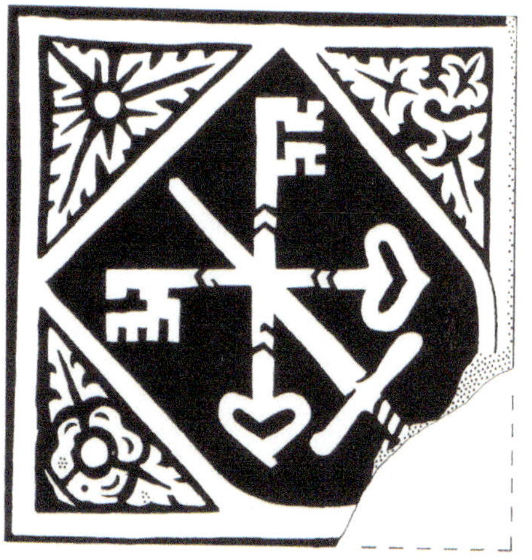 471
The Westhide example has the sword pointing downwards towards the handles of the keys, and the keys themselves face each other unlike the Ewenny tile opposite. Also the same three-petalled rose and leaves fill the three corners, unlike the Ewenny Priory example. The stamp appears too big for the tile.	JM Lewis Group 26 (Ewenny Priory tiles) 471. Shield of arms set diagonally; *crossed keys a sword in pale* (Gloucester Abbey); leaf motifs above and to right, to left a three-petalled rose and leaves; the whole framed. (Early 16th century?).

Wigmore Abbey

An Augustinian abbey was founded at Adforton, near Wigmore, in 1179 by Sir Hugh de Mortimer. It was the largest in the county of Herefordshire, followed by Dore Abbey and Leominster. There are inevitable close links with Wigmore Castle, where the Woolhope Group visited in 1870 and which is recorded in their Transactions of that year.

'. . .Wigmore Castle: In excavating the steps into a vault in the basement of one of the angle towers to the left of the entrance gateway some fragments of stags horn, belts, hinges and an old lock were found, and among the tiles exhumed was one bearing the old royal arms, gules, three lions passant gardant in pale or.'

These tiles were not curated.

I understand there have been several archaeological excavations at Wigmore Abbey, including the excavation of the chapel floor during a BBC TV programme called The House Detectives, but I can find nothing published from these digs. I believe Alan Hunt excavated here in 1975. The most informative drawing was published by H. Brakspear in the Archaeological Journal 90 in 1933 which shows a 16-tile design from the Presbytery of Wigmore Abbey.

The abbey was of course a victim of Henry VIII's Dissolution and is now in ruins but part of it was incorporated into a large medieval farmhouse which is now in private ownership. The owner very kindly allowed access to the boxes of tile fragments which he had acquired with the house.

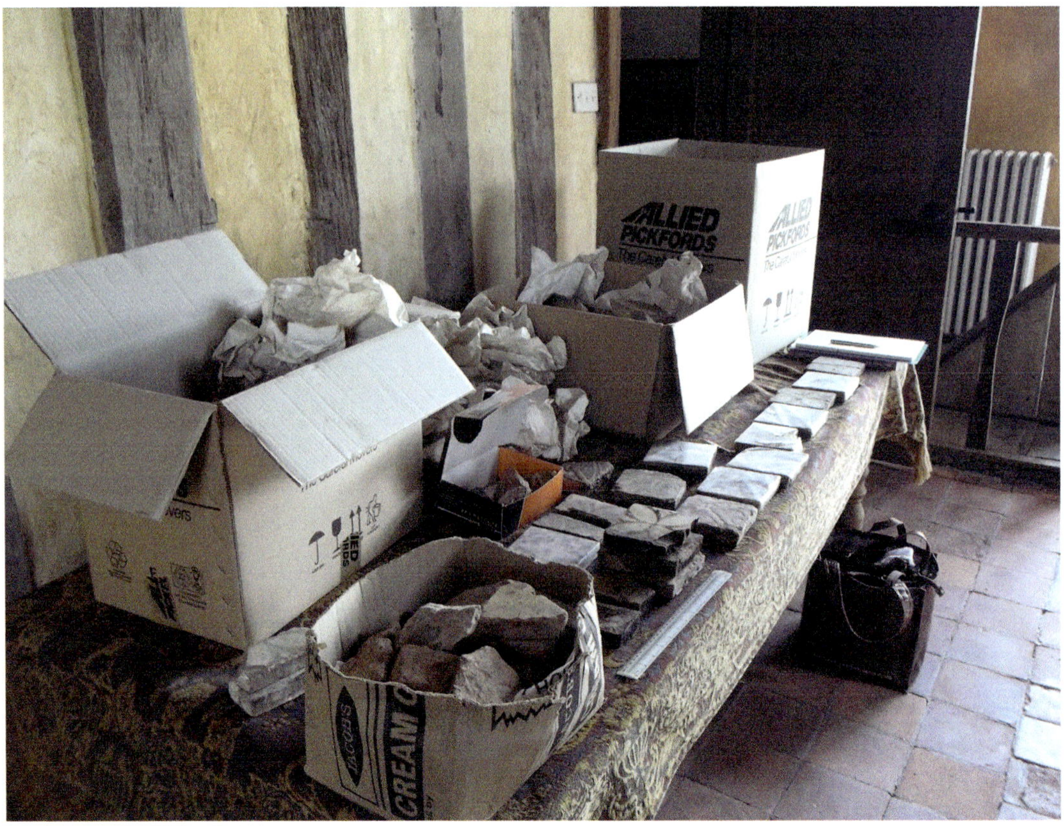

The majority are the counter-relief tiles referred to in JM Lewis's book as Group 27 and which are found in Wales only at Old Radnor Church (close to Wigmore).

These same counter-relief designs have also been found at Chapel House, Wigmore, at Oakley House, Wigmore, and at Brampton Bryan church. Wigmore Castle and Brampton Bryan castle were both owned by the Harley family, who at the time of the Civil War deliberately destroyed Wigmore Castle to prevent its use by Royalists.

The most common tiles are from the Brakspear design, some with a bright yellow glaze, and also with dark green, and dark brown/black glazes.

This drawing was published by Harold Brakspear in his article on Wigmore Abbey in The Archaeological Journal of 1933.

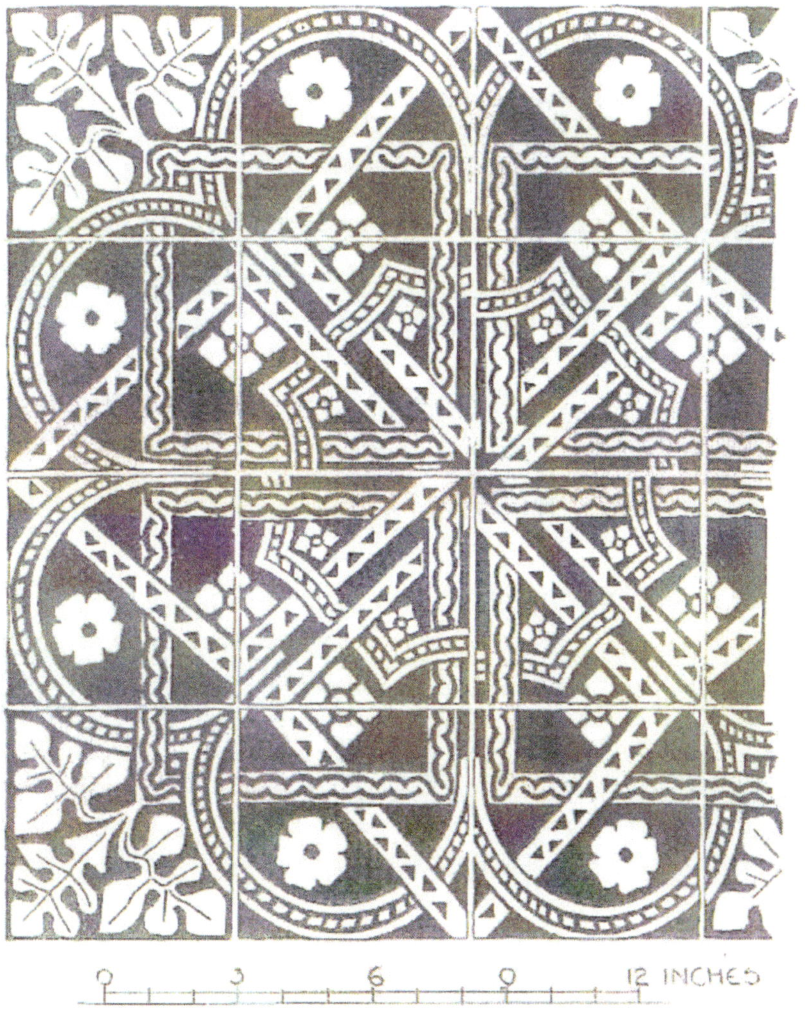

WIGMORE ABBEY : TILES FROM THE PRESBYTERY

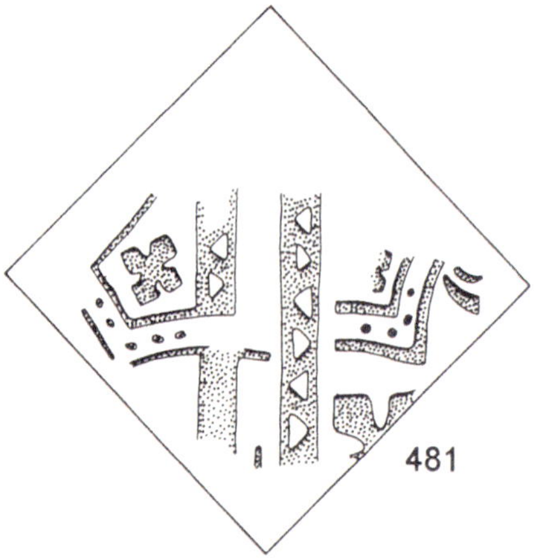	481 JM Lewis. Relief: 4-tile design (or larger); mitre shape set diagonally with interwoven bands. In the Presbytery at Wigmore Abbey it forms the centre of a 16-tile design (Brakspear 1933, 33 and pl.vi). Herefordshire: Wigmore Abbey; Chapel Farm, Wigmore; Oakley House, Wigmore; Brampton Bryan Also at: Old Radnor Church.

Examples of the different coloured glazes:

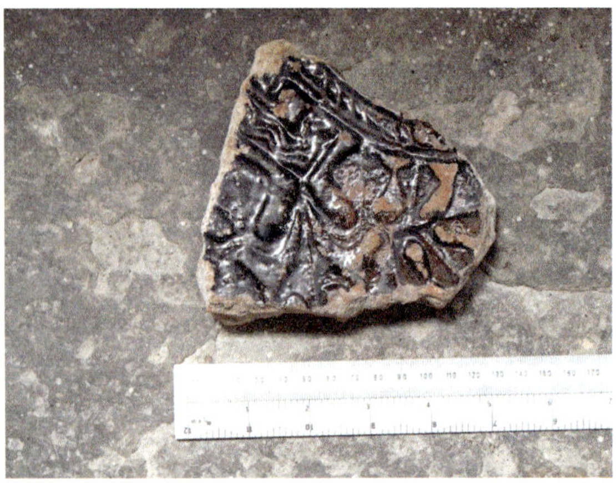

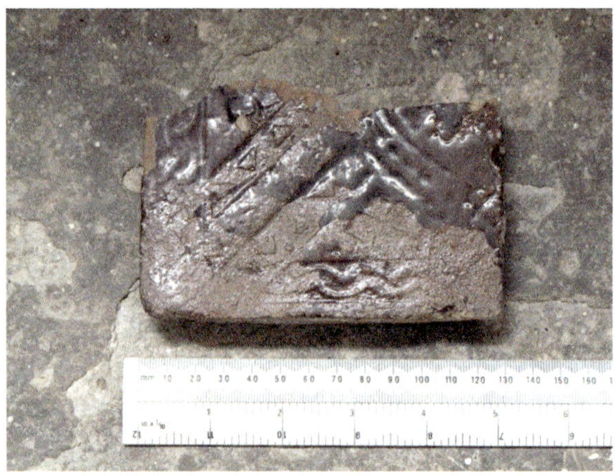

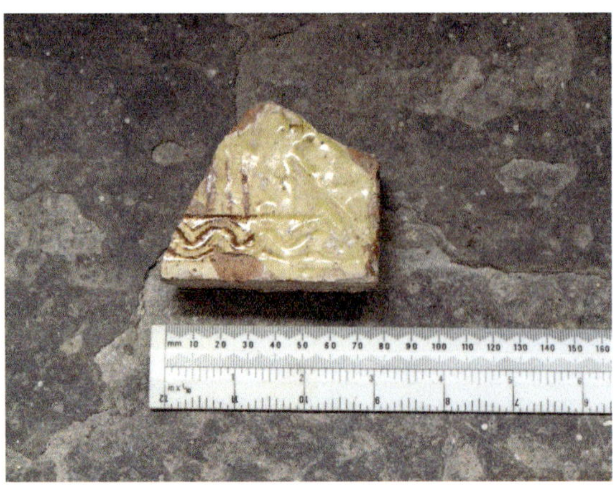

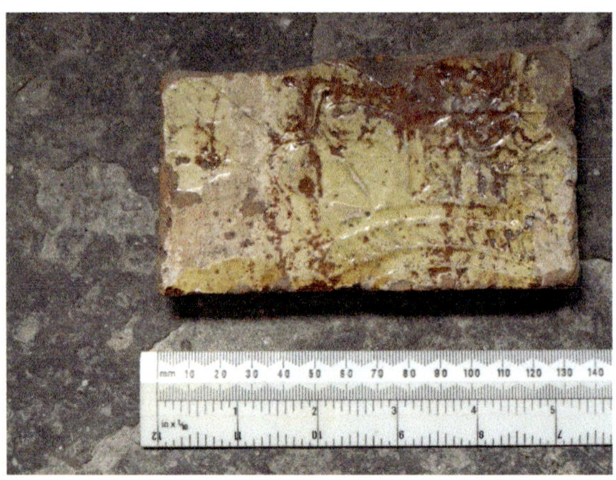

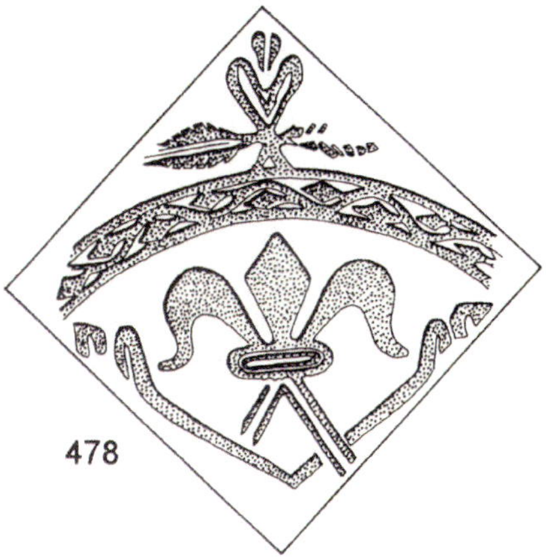

478

478 JM Lewis.
Counter-relief: 4-tile design; circular band with running leaf design surrounds a cross; in each interspace a fleur-de-lys connected to a double-stranded ribbon interwoven with the arms of the cross; in each corner outside the circle a plant motif.

At Wigmore Abbey examples with a yellow glaze and dark green glaze.

Herefordshire:
Chapel Farm, Wigmore

Also at: Old Radnor Church.

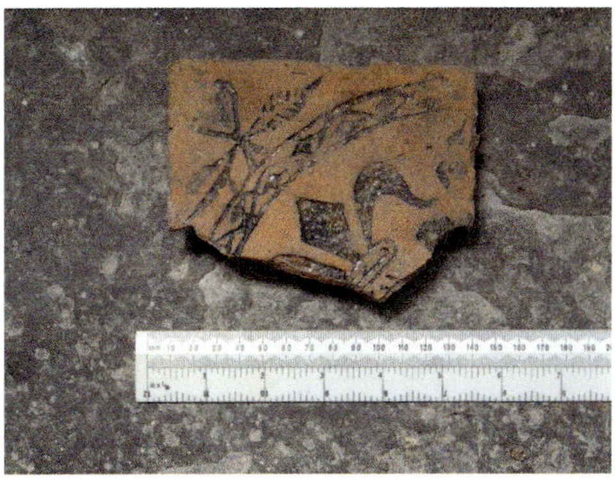

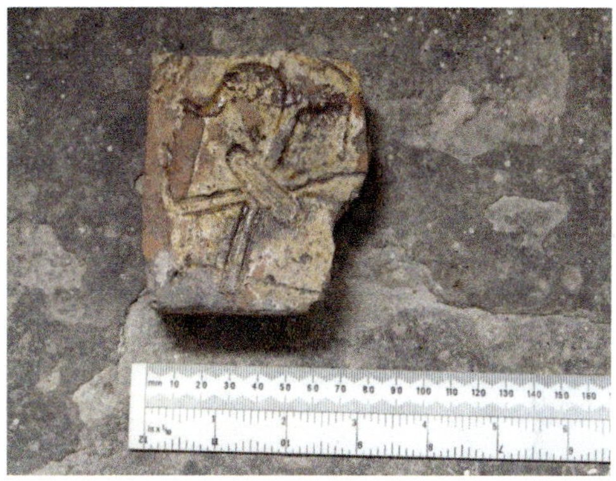

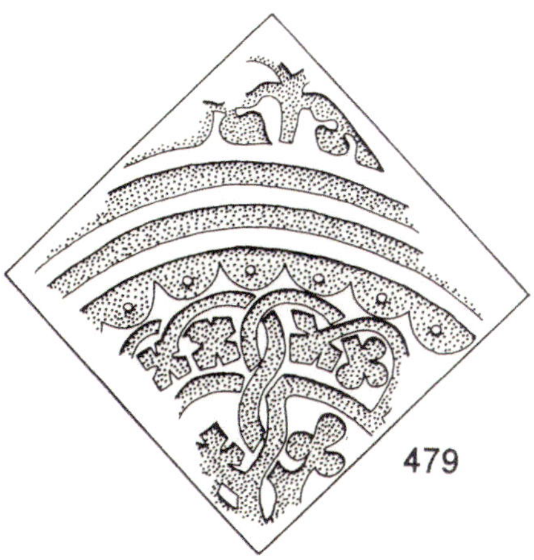

479

479 JM Lewis.
Counter-relief; 4-tile design; double circle surrounding an interwoven leaf-and-tendril design.

At Wigmore examples with a dark green glaze.

Herefordshire:
Chapel Farm, Wigmore

Also at: Old Radnor Church.

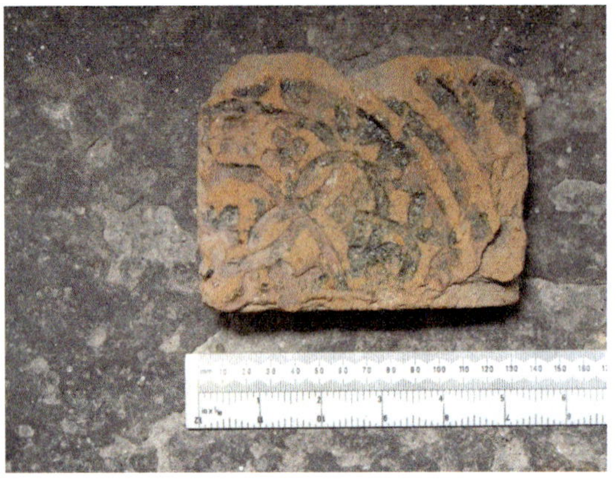

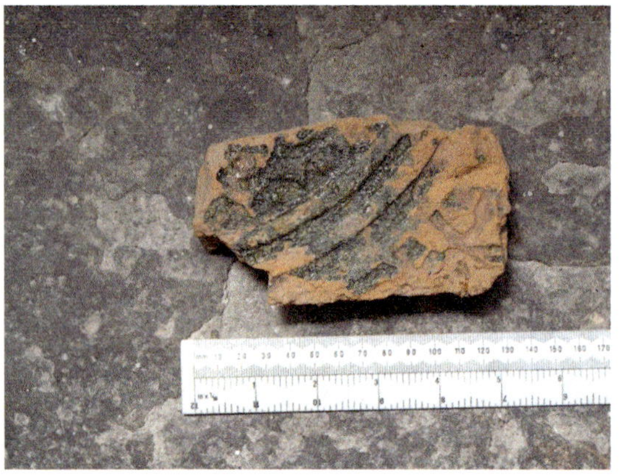

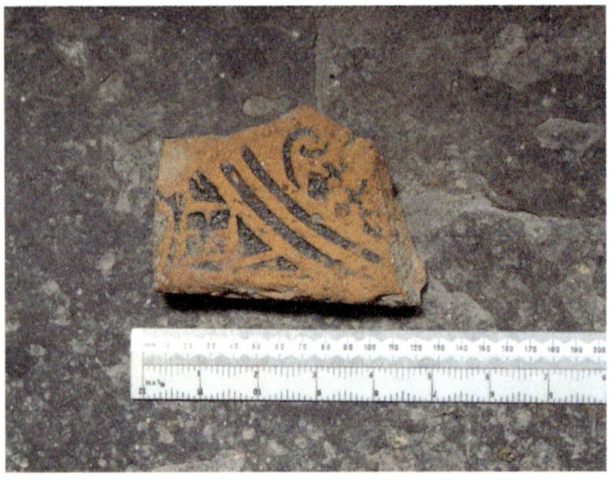

This tile (left) appears to have a lozenge in the corner as opposed to the weeping foliage depicted above and in the JM Lewis drawing.

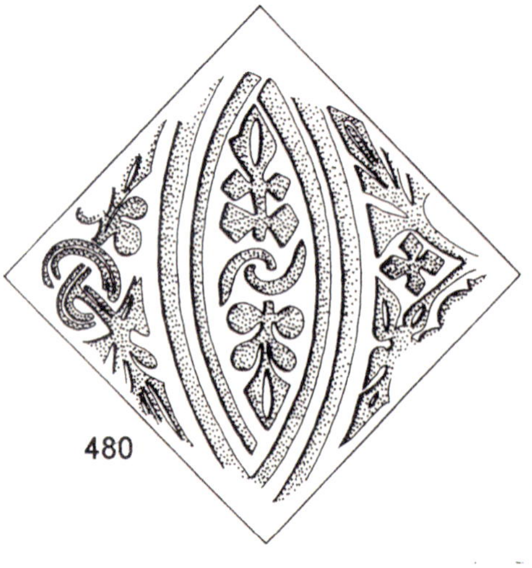

480

480 JM Lewis.
Counter-relief; carpet pattern; double-outline vessica set diagonally, forming interlinked circles containing leaf designs.

At Wigmore Abbey examples in a yellow, dark green, and black glaze. (See below.)

Herefordshire:
Chapel Farm, Wigmore; Brampton Bryan.

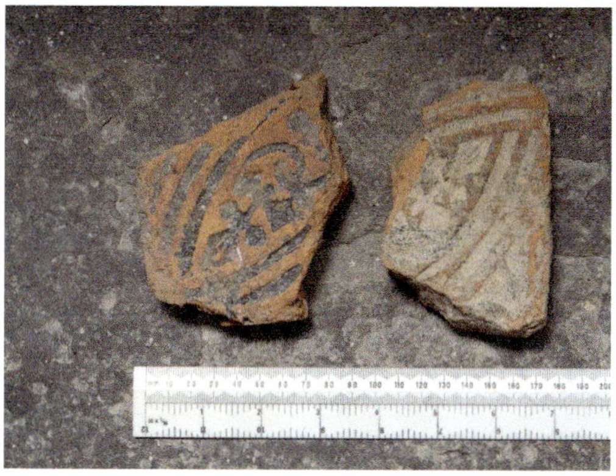

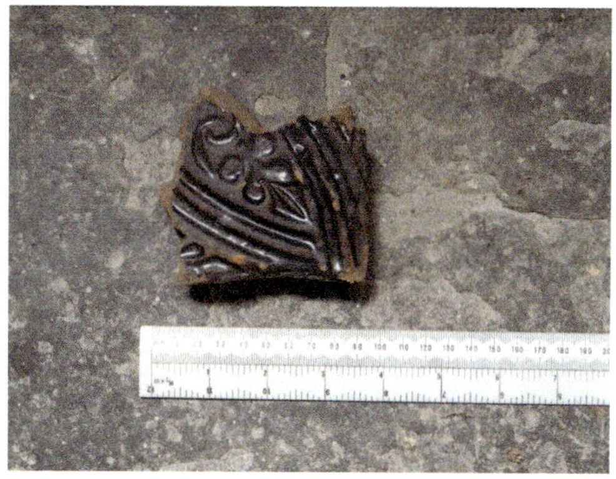

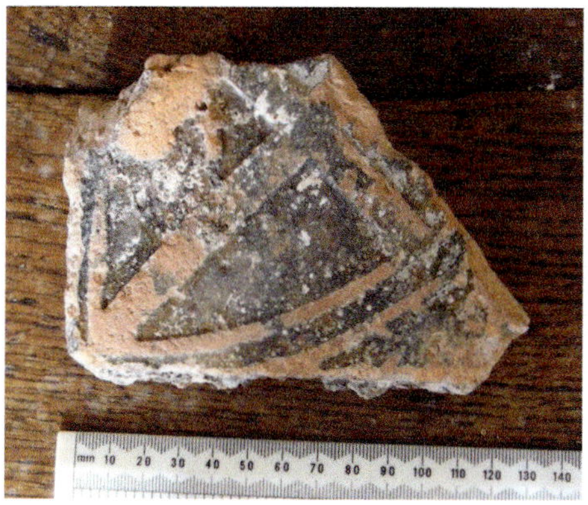

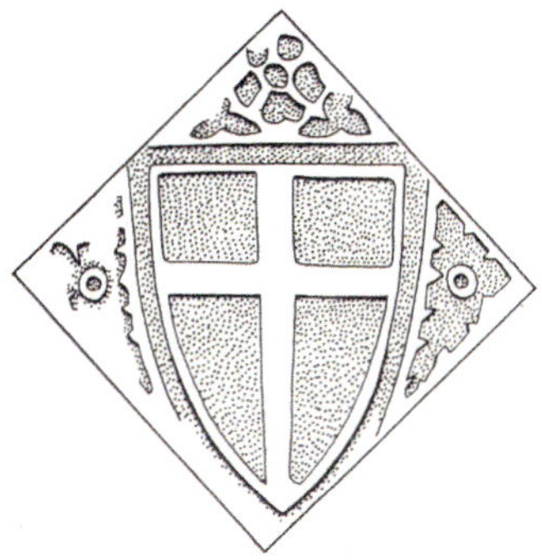

477

477 JM Lewis.
Relief/counter-relief design. Heraldic; a cross set diagonally; flower and leaf motifs above and at the sides.

Also at Old Radnor Church.

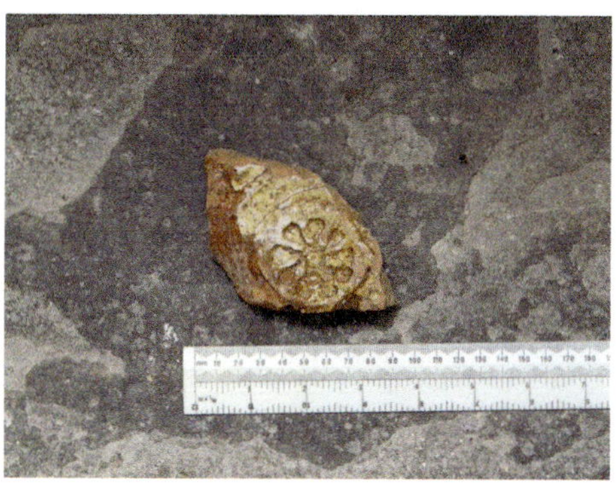

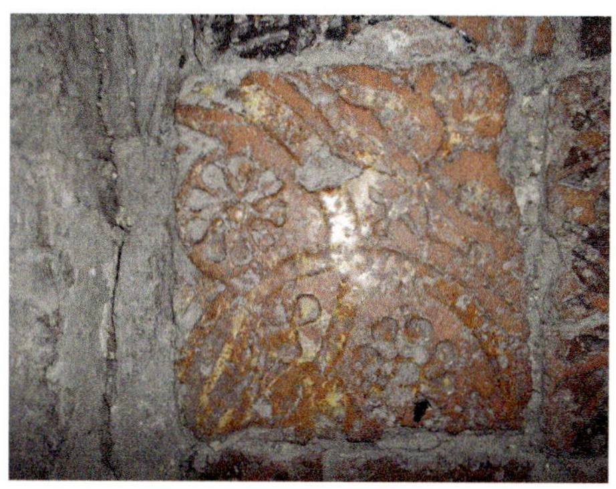

This tile (left) from Wigmore Abbey exhibits the same wheel found on a tile (right) from Chapel Farm, Wigmore, but I can find no parallels for the design.

The following four designs are also found in the Lady Chapel at Hereford Cathedral, and Worcester Cathedral.

Wigmore Abbey tiles

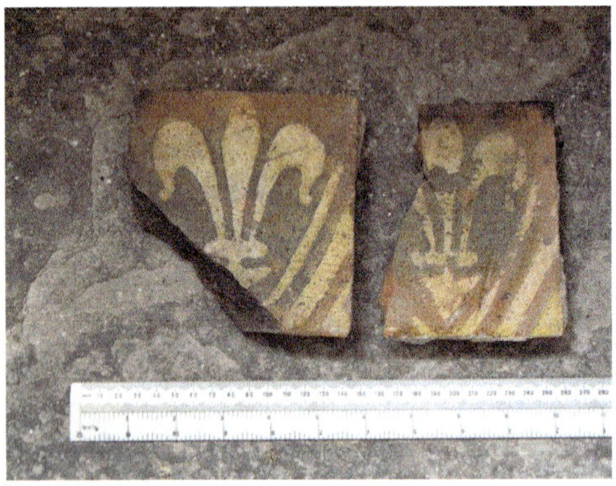

42

This similar design is from Worcester Cathedral, (and the same design is found at Hereford Cathedral, Lady Chapel) but the Wigmore example has a double outline to the shield, different corner decoration, and a different base to the fleur-de-lys.

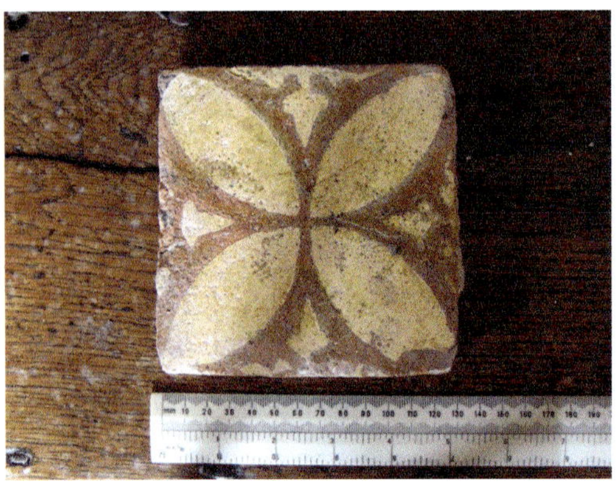

Also at Hereford Cathedral, Lady Chapel.

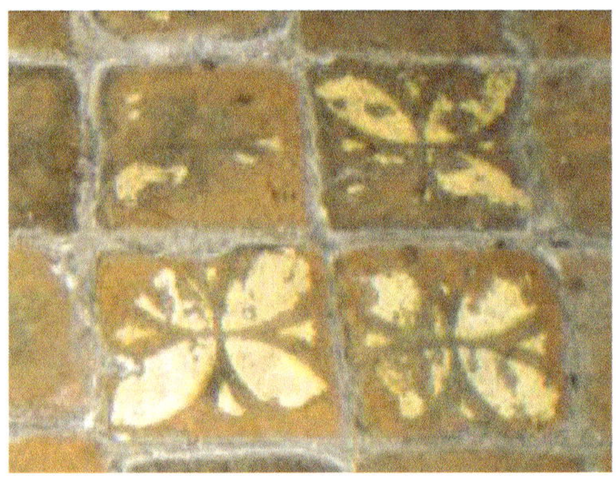

Laurence Keen photo

Similar at Worcester Cathedral.

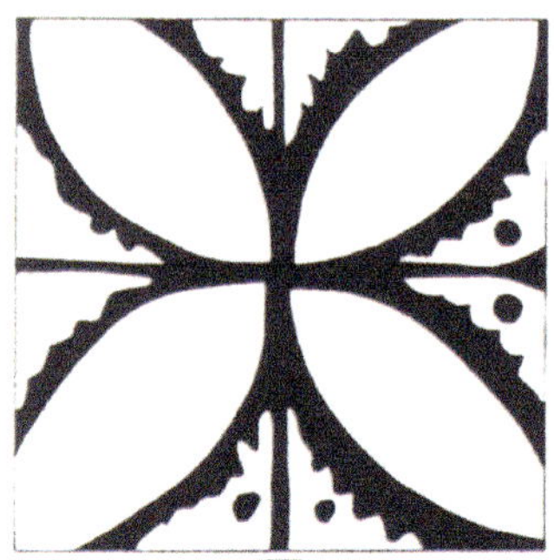

17

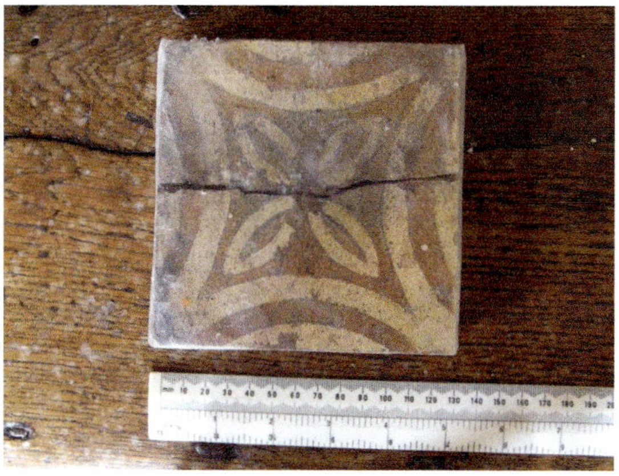

Also at Hereford Cathedral, Lady Chapel.

29

Worcester Cathedral.

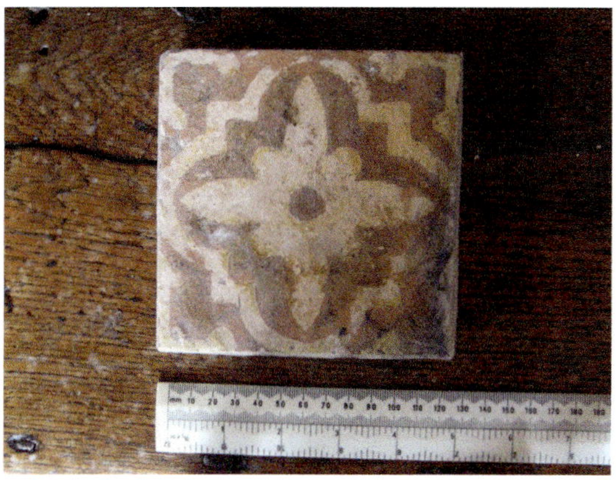

Also at Hereford Cathedral, Lady Chapel

28

Worcester Cathedral.

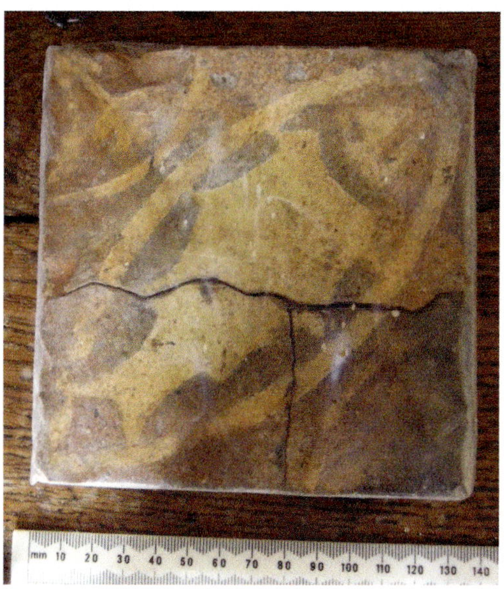

Worcester Cathedral.

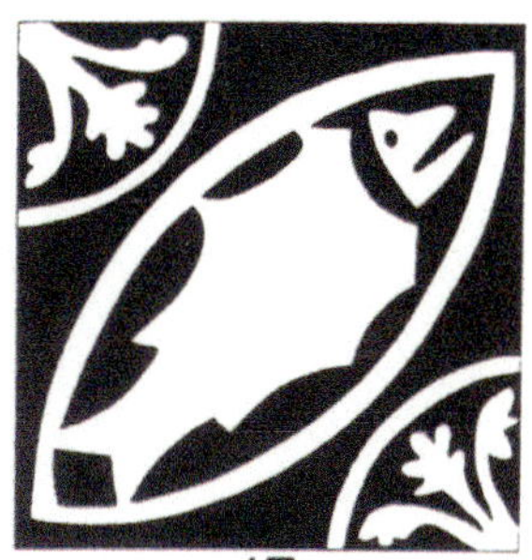

15

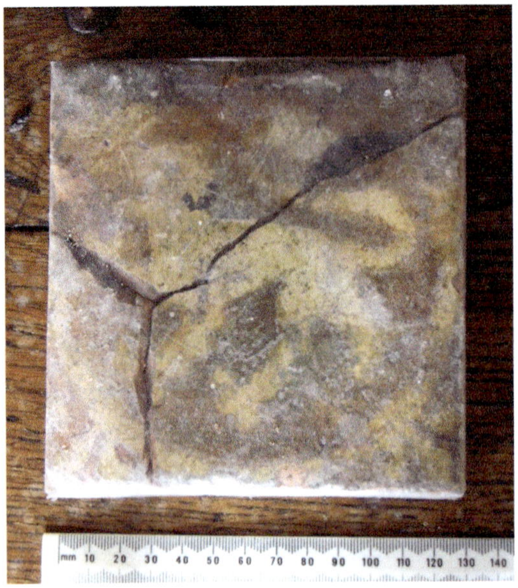

JM Lewis 34.

Also see Netherwood, Thornbury.

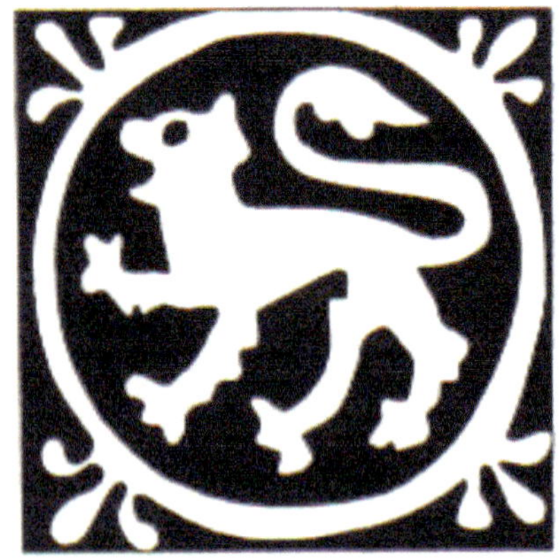

Worcester Cathedral.

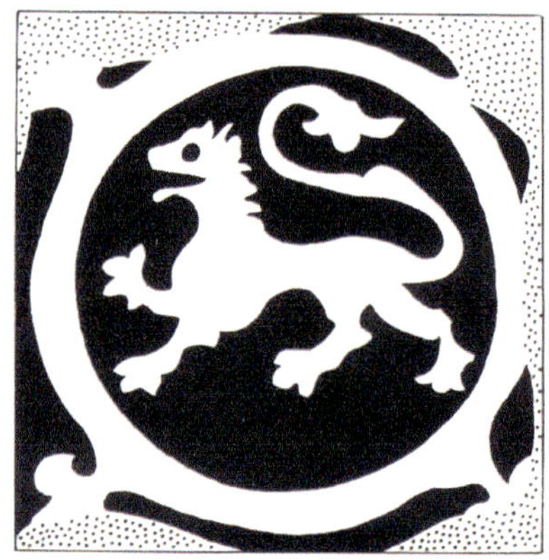

34

JM Lewis Group 8.
34
Inlaid. Lion passant, within circle, with fleurs de lys at corners.
BM Design 1783.

Also at Monmouth (x2), Goodrich Castle.

Second half of 13th century.

<table>
<tr>
<td>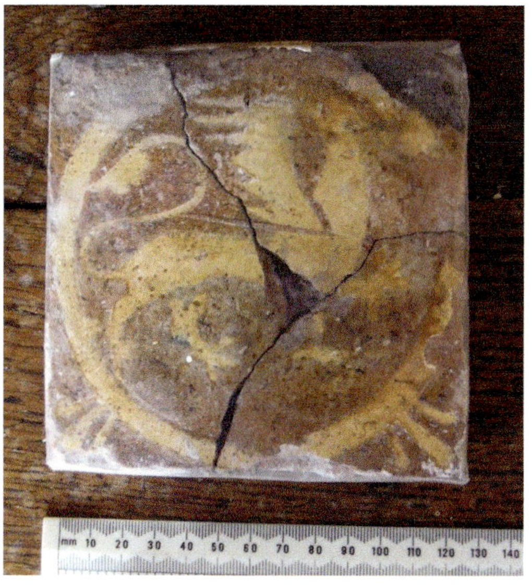

Finely drawn – likely to be early.
(the wings and corners are different from the JM Lewis example opposite).</td>
<td>JM Lewis 33
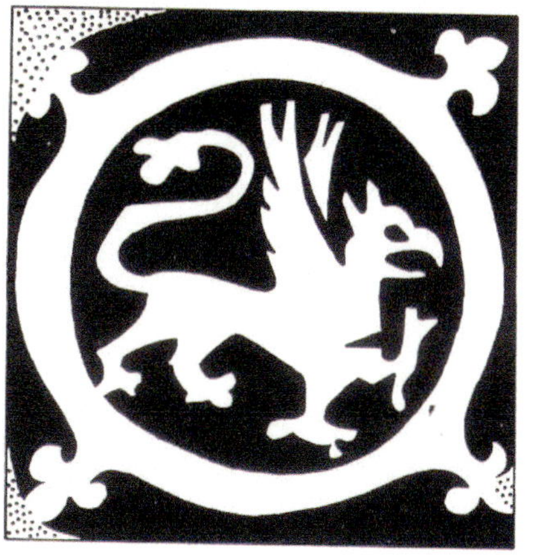
33
JM Lewis Group 8.
33
Inlaid. Griffin passant contourne, within circle, with fleurs de lys at the corners.
(Wessex series?).

Also at Monmouth (x2), Goodrich Castle.

Second half of 13th century.</td>
</tr>
<tr>
<td>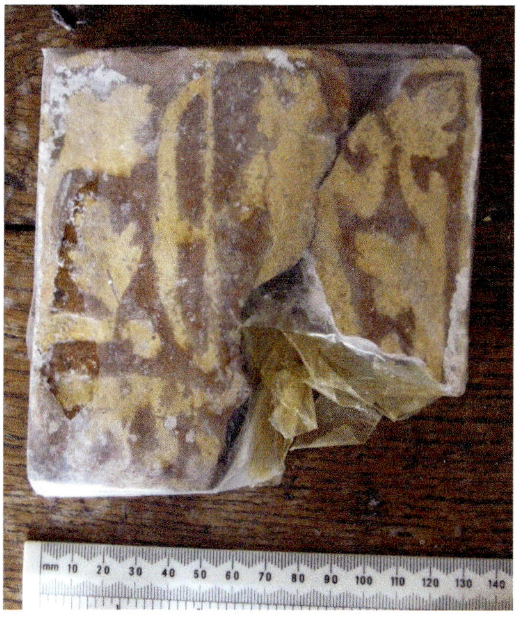
</td>
<td>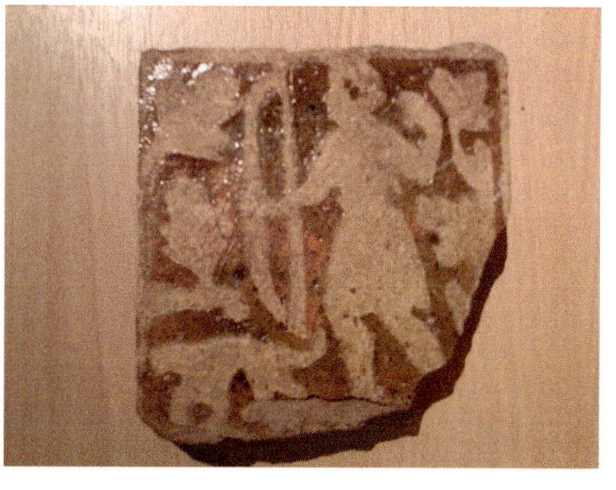
From Greyfriars, Hereford
(in Hereford Museum).

Archer with dog (bottom left).</td>
</tr>
</table>

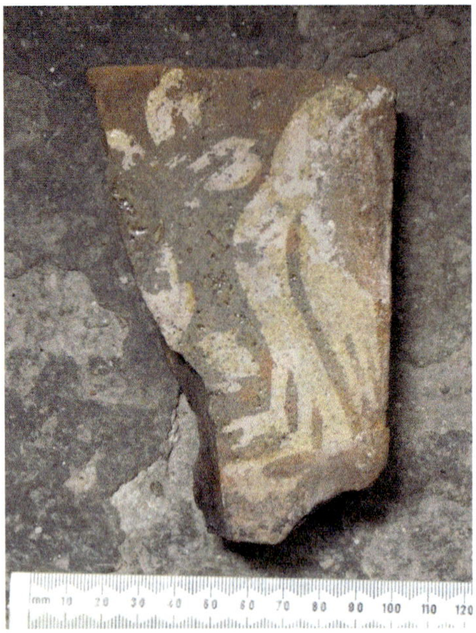

c.f. Worcester Cathedral 18.

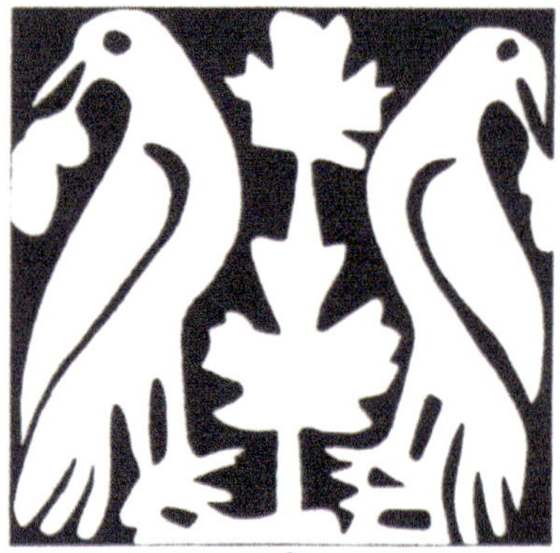

18

Head looking up in Wigmore tile (?) and different central foliage.

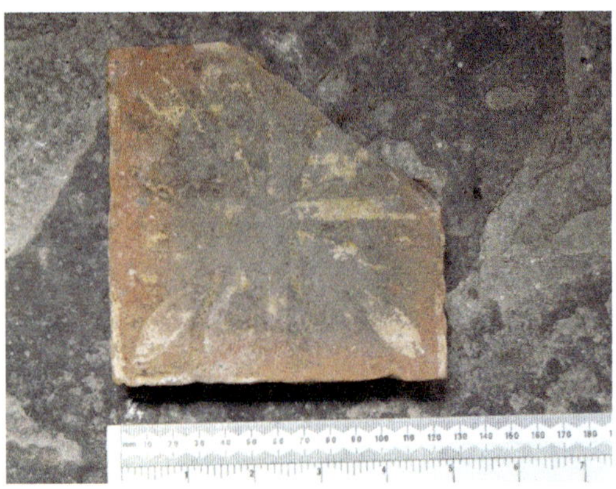

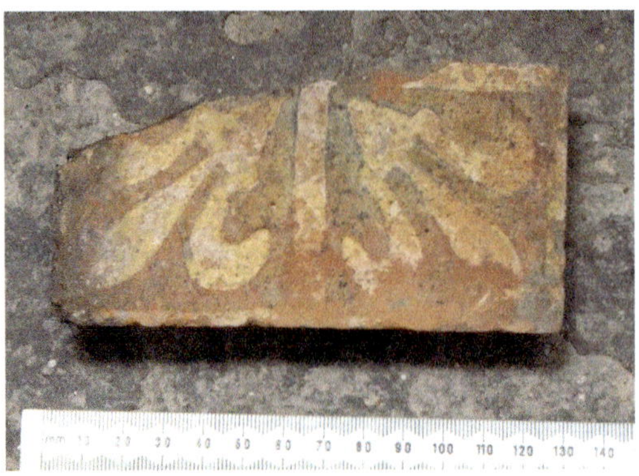

27

Worcester Cathedral
Similar to BM 2205.

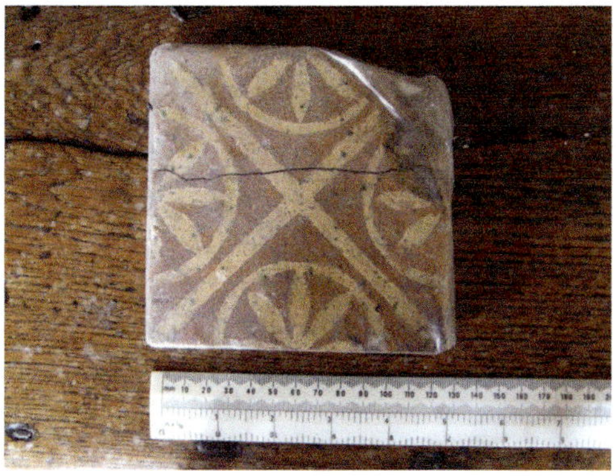

After Eames: BM 2446.
Maxstoke Priory, West Midlands c. 1448.

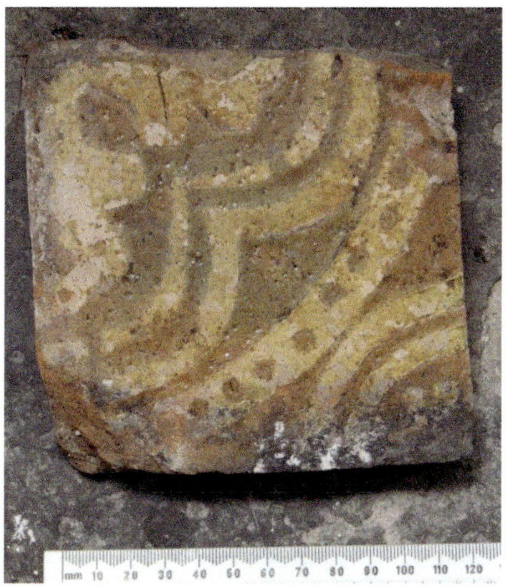

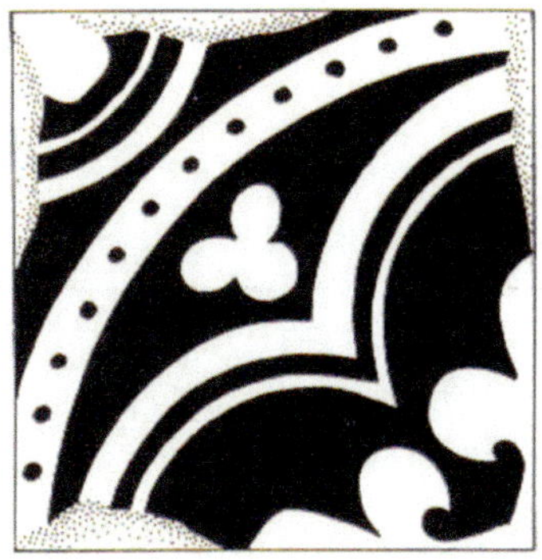

81

This design from Worcester Cathedral is similar but the Wigmore tile does not have the tri-lobed feature.

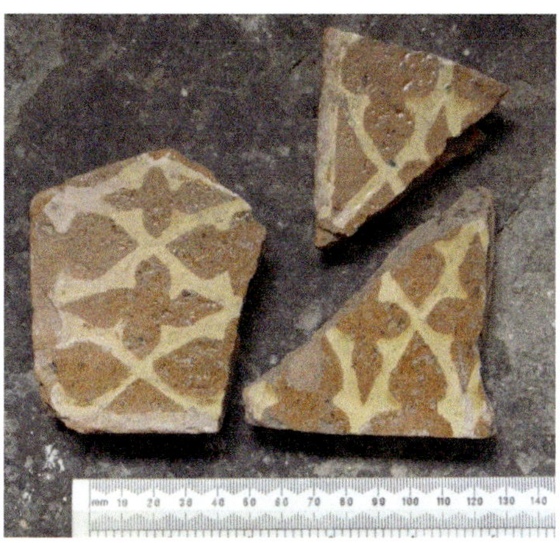

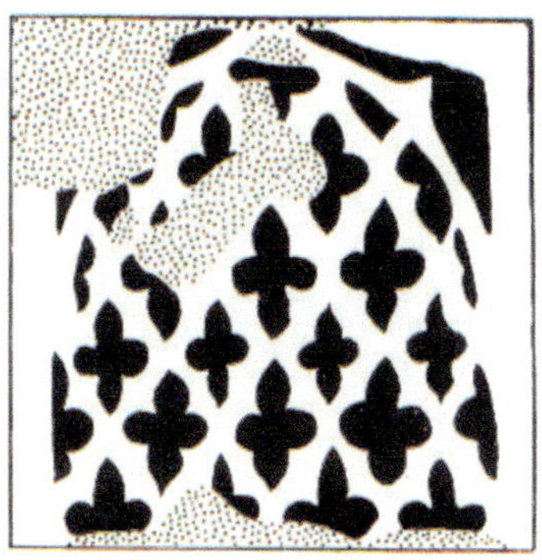

152

JM Lewis Slip on impression. Window head with quatrefoil tracery (Neath Abbey).

The following examples are all unidentifiable at this stage.

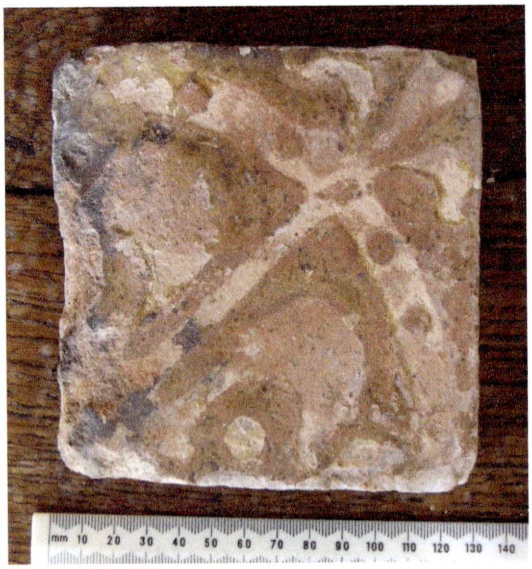

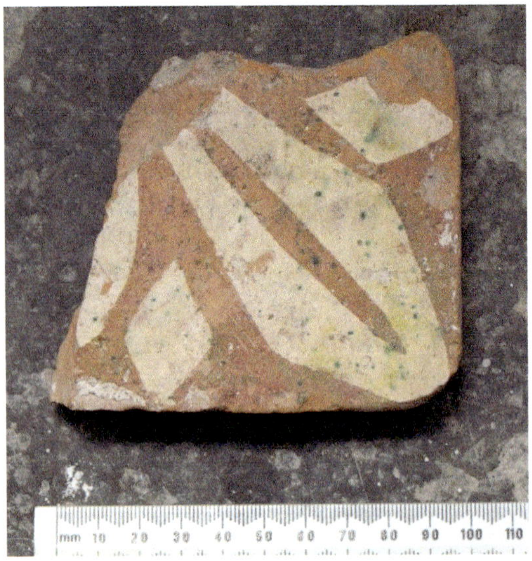

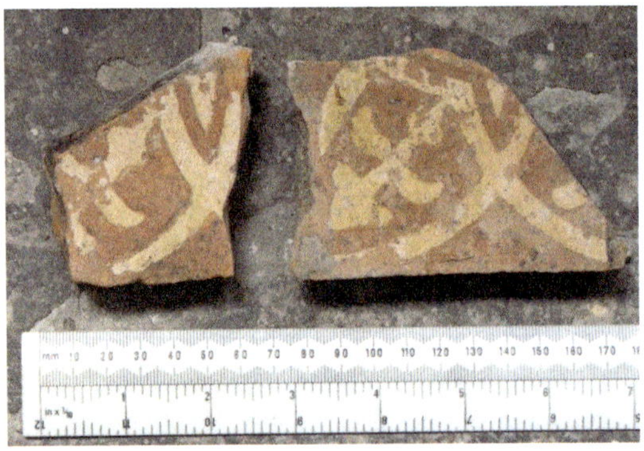

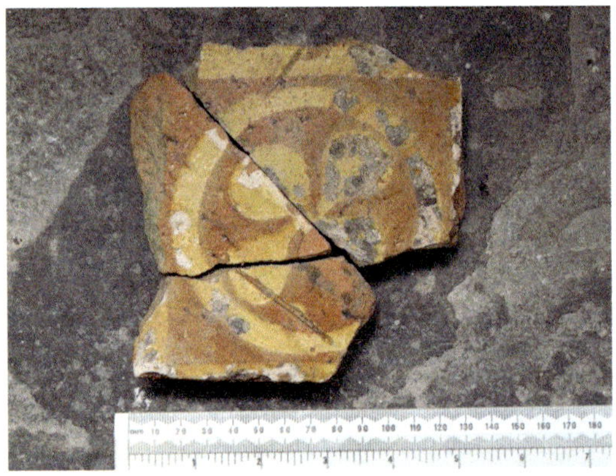

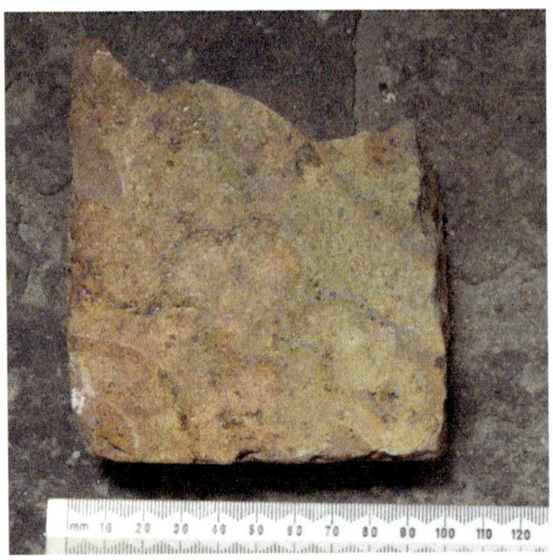

These two examples of tiles fused together during firing would suggest the kiln site was close by to Wigmore Abbey.

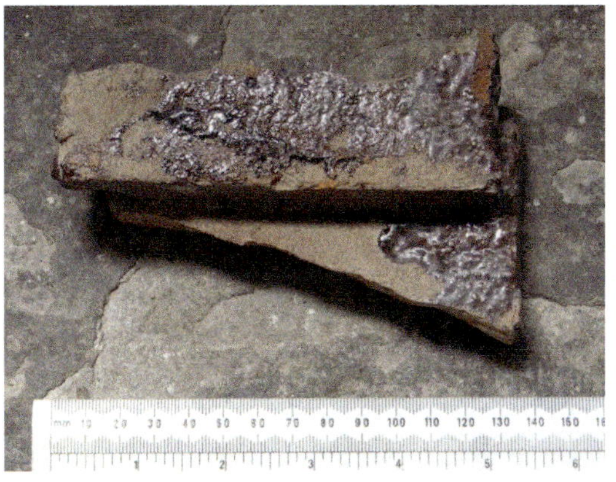

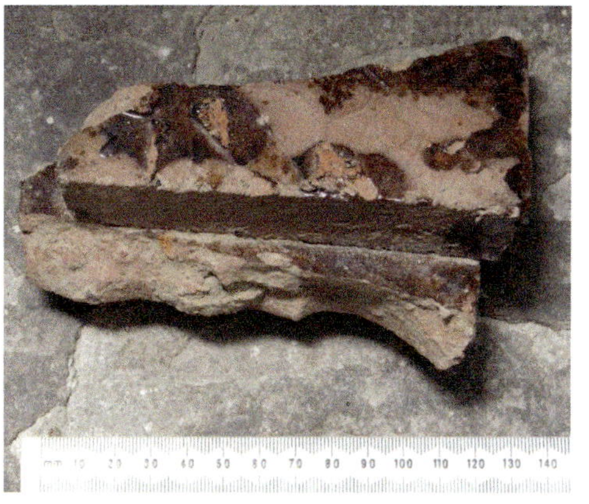

In the collections of the Hereford Museum there are several fragments of plain glazed floortile from a 1975 excavation at Wigmore Abbey. These include a dark green/brown lozenge shape measuring approx. 18cm by 16cm; a black glazed square tile 11.7cm, and a yellow glazed triangle 11.7cm.

Wigmore – Chapel Farm

Chapel Farm is a beautifully preserved medieval farmhouse in private ownership. It is reputed to be the site of the chapel of the Blessed St Mary and St Leonard which was given to the nuns of the nearby priory of Limebrook by Roger de Mortimer of Wigmore in 1250. The present building dates from the late 15th century but is thought to be on the site of an earlier structure. The medieval floortiles, which are believed to have come from the chantry, have been relaid as a hearth in a first floor bedroom.

They are roughly laid and many have cracked in situ.

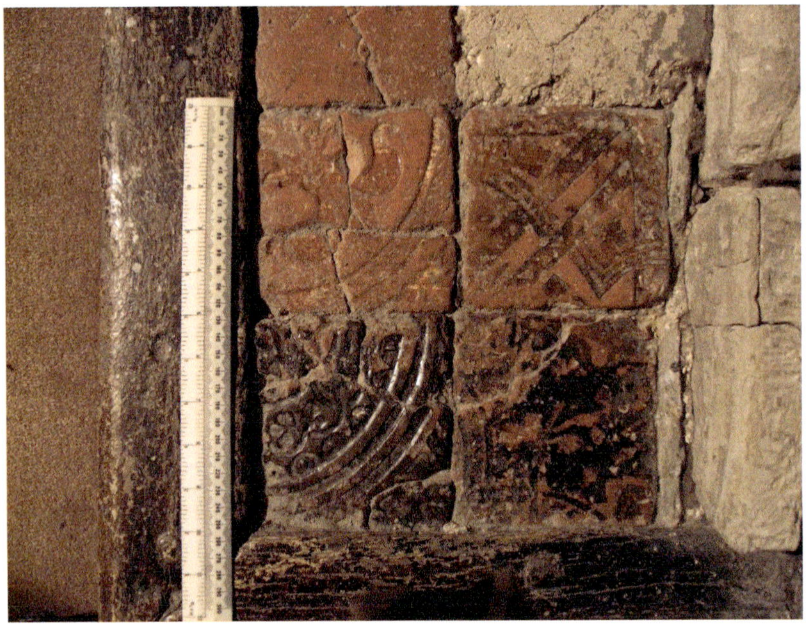

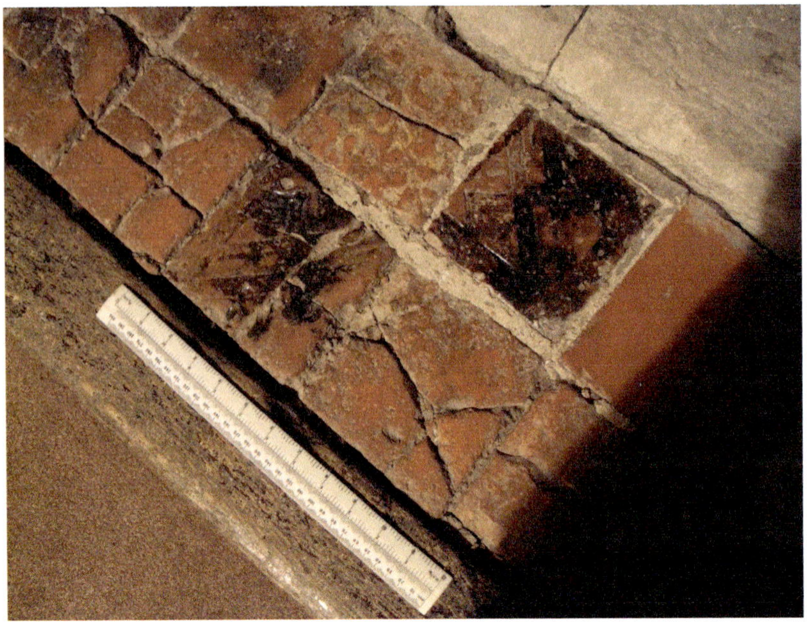

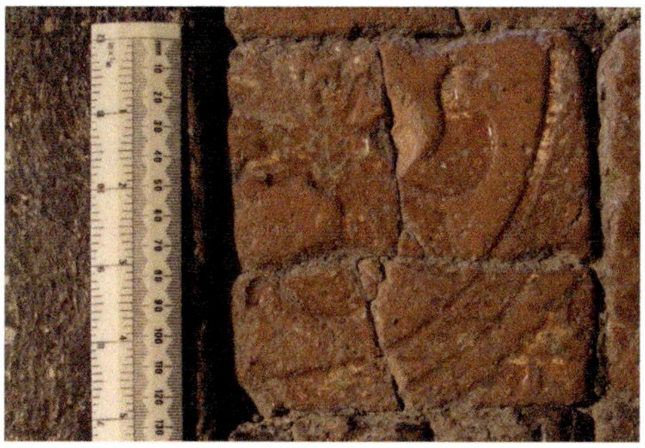

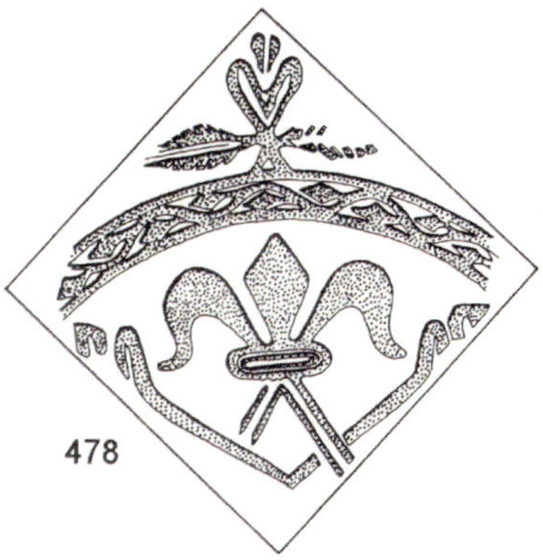

478

Group 27 of JM Lewis.
478
Counter-relief: 4-tile design; circular band with running leaf design surrounds a cross; in each interspace a fleur-de-lys connected to a double-stranded ribbon interwoven with the arms of the cross; in each corner outside the circle a plant motif.

A single example in an orange glaze at Chapel Farm, Wigmore.

Also at Wigmore Abbey, and Old Radnor Church.

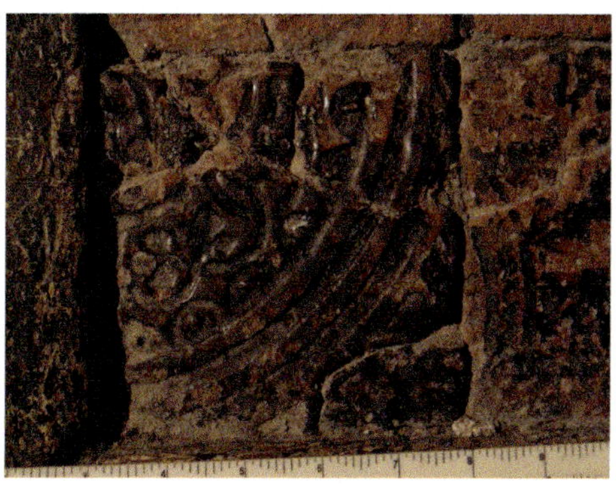

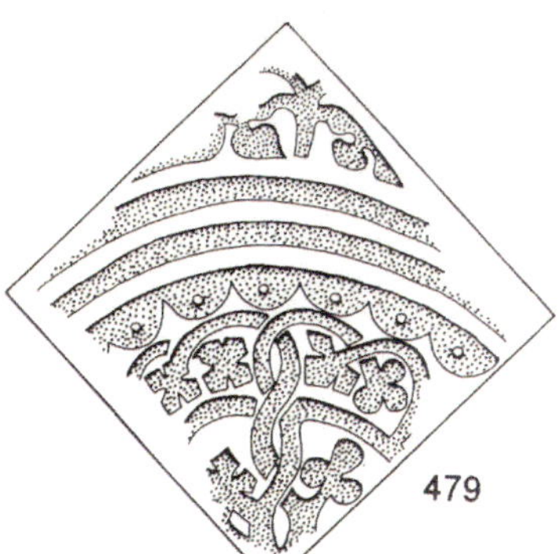

479

479
Counter-relief; 4-tile design; double circle surrounding an interwoven leaf-and-tendril design.

A single example in a black glaze at Chapel Farm, Wigmore.

Also at: Wigmore Abbey, Old Radnor Church.

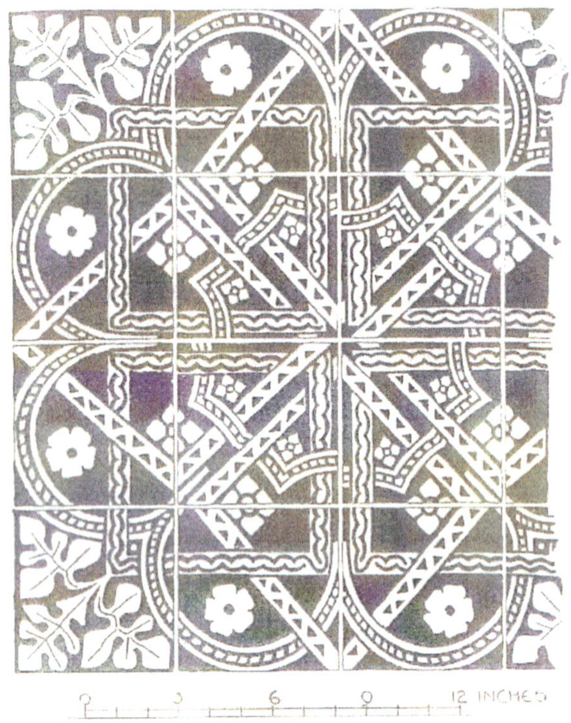

WIGMORE ABBEY : TILES FROM THE PRESBYTERY

481
Relief: 16-tile design; mitre shape set diagonally with interwoven bands. In the Presbytery at Wigmore Abbey it forms a 16-tile design (Brakspear 1933, 33 and pl.vi).

Herefordshire:
Wigmore Abbey, Brampton Bryan.

Also at: Old Radnor Church.

Tiles from this design are represented at Chapel Farm with a yellow, brown, dark green, and black glaze. There are more examples of the corner tiles displaying the three leaves than of the inner tiles.

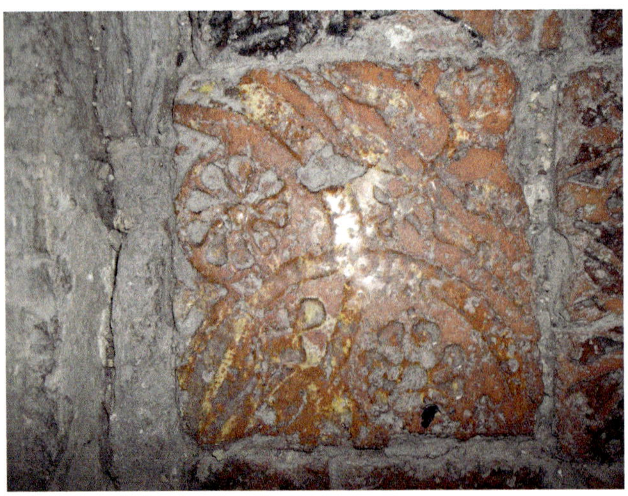

A relief design with a yellow glaze. No parallels found.

A small fragment from Wigmore Abbey is probably the same design.

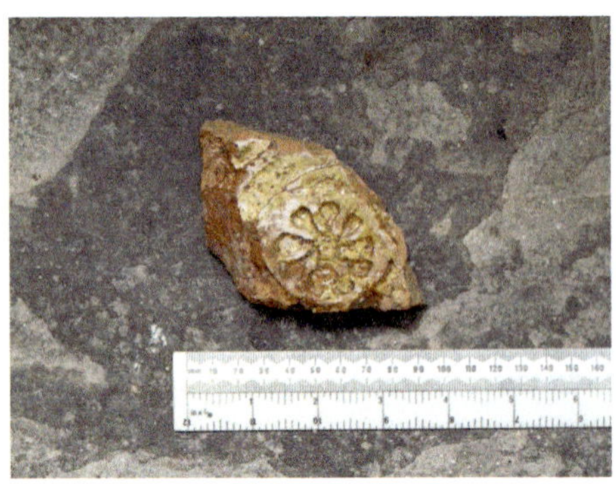

Yatton – Deans Place

Deans Place is a beautifully preserved medieval farmhouse, now in private ownership. In a first-floor bedroom are 18 medieval slip-on-impression floortiles, originally used as a hearth but now re-set above the mantelpiece, and until recently covered by wallpaper. Showing little signs of wear, the tiles measure approx 120cm square, and are laid in two rows of nine tiles. Despite the name, Deans Place has no record of an ecclesiastical history, but John Clark, a local historian, believes they may have come from a Norman chapel in the parish which was abandoned in 1841. All of the tiles carry designs from the Canynges Pavement in Bristol, which are also found nearby at St Katherine's Chapel in Ledbury with a likely date around 1485.

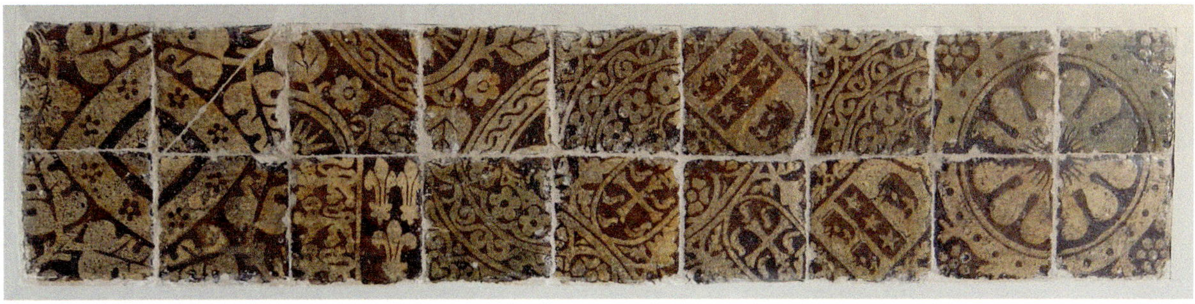

(photo courtesy of Dr Patrick Brown)

At each end is a different group of four tiles. Dr Patrick Brown has identified the left hand group from the British Museum collection of the Canynges Pavement.

The group of four tiles on the right form the centre of a 16-tile Bristol Canynges Pavement Design VII (BM 2991-3) (JM Lewis 383-6) also found at St Katherine's Chapel, Ledbury, Gloucester Cathedral Lady Chapel, and at Shelsley Walsh (Worcs):

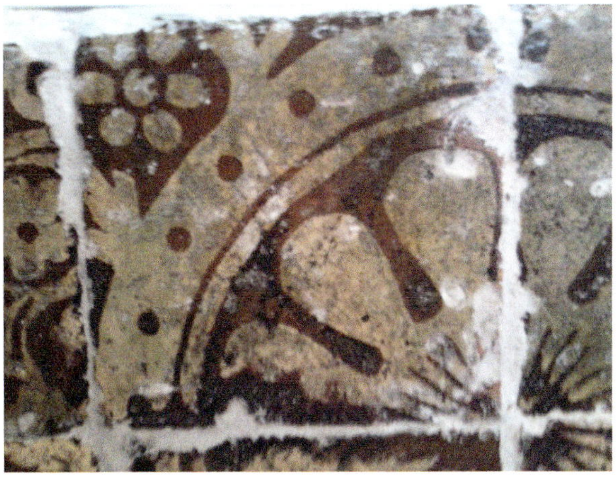

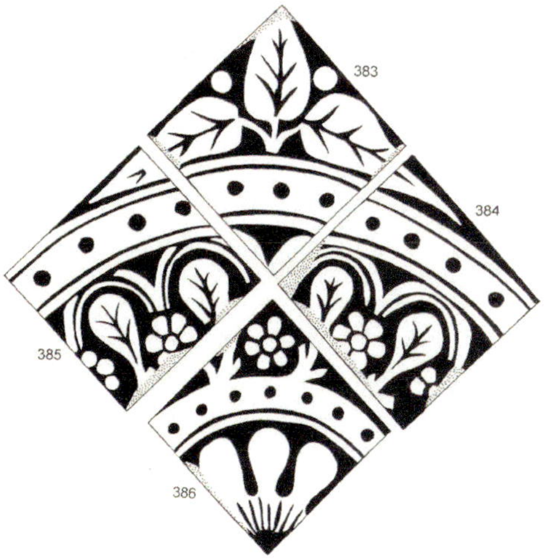

'16-tile design; circular band with reserved pellets surrounding a frieze of arcading, each arch containing a leaf that springs from an inner circle (also with pellets), between each pair of leaves a six-petalled rosette; in the centre a twelve-petalled flower; in each corner outside the circle three lanceolate leaves with pellets between.'

Other Canynges designs are evidenced in Canynges IV (BM Design 2972 - JM Lewis 403) below. Three of these tiles are present, one of which shows a diagonal score (top left to bottom right):

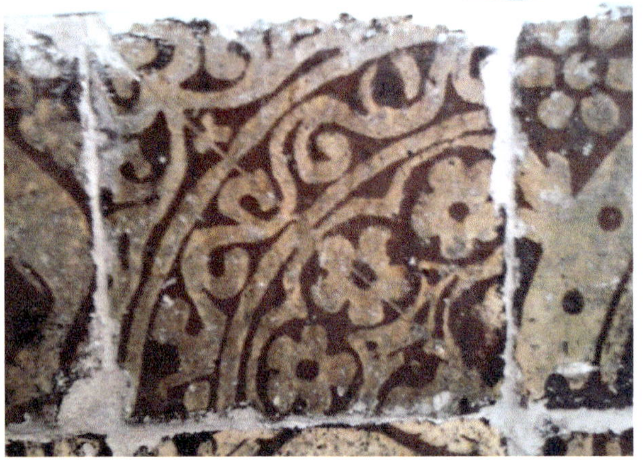

'4-tile design: circular band of scroll design enclosing a band of twelve five-petalled rosettes, with a further eight at the centre.'

This design has also been found at St Oswald's Armhouses, Worcester and at Shelsley Walsh (Worcs).

Another Canynges Design I is present (BM 299 / JM Lewis 367-9), with two examples of tile 367:

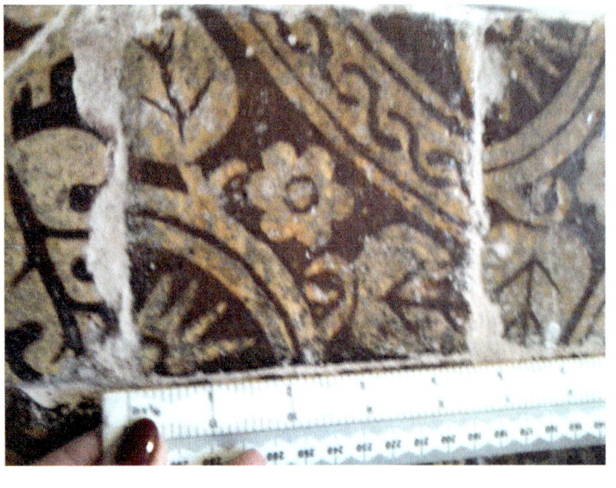

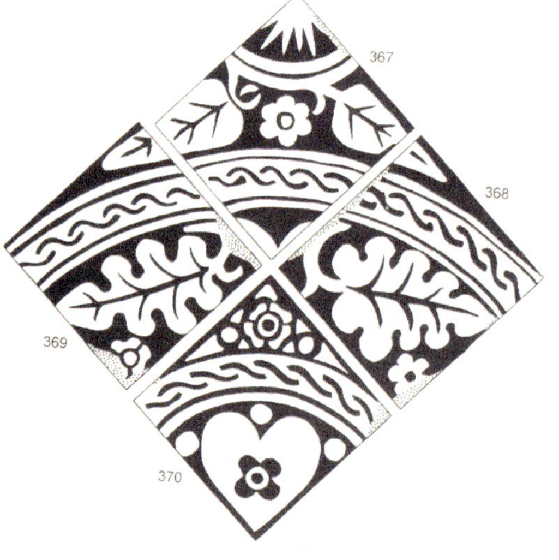

'16-tile design; circular band with simple cable pattern enclosing oak leaves; the circular band is repeated at the centre, where it encloses a heart pierced with a four petalled rosette; in each corner outside the larger circle a six petalled rosette flanked by leaves springing from a double circle surrounding a sun-burst.'

This is also found at St Katherine's, Ledbury.

There are two examples of the corner tiles from another Canynges Design IX (BM 2893 / JM Lewis 395-8):

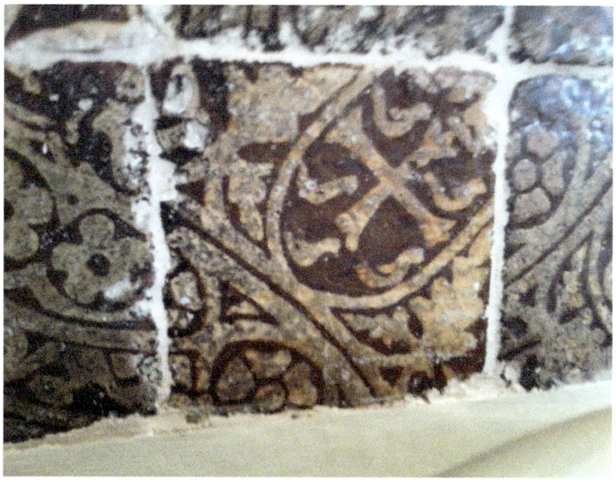

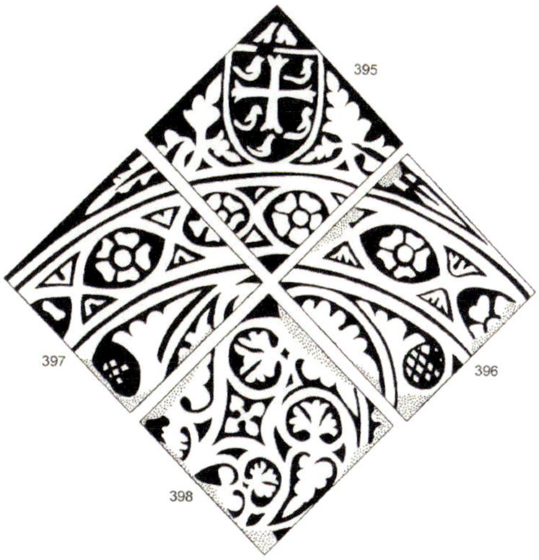

'16-tile design; circular band with five-petalled rosettes in a cable pattern surrounding a quatrefoil containing an asymmetrical foliage design; in each corner outside the circle a shield of arms; *between five martlets contournee a cross patonce* (Westminster Abbey reversed).'

There are two tiles showing the heraldic shield of Wyatt of Tewkesbury (JM Lewis 453):

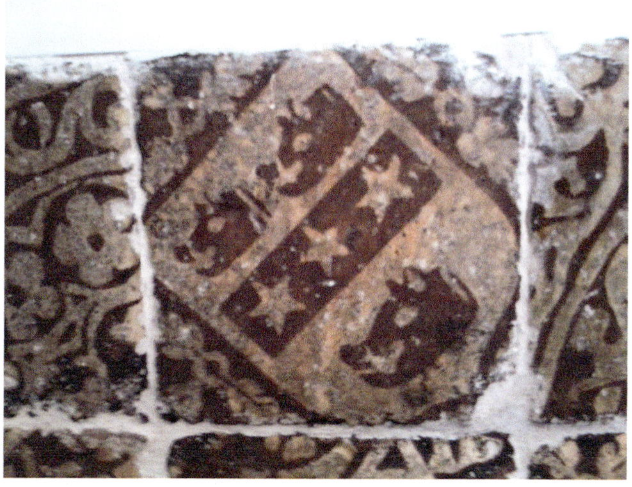

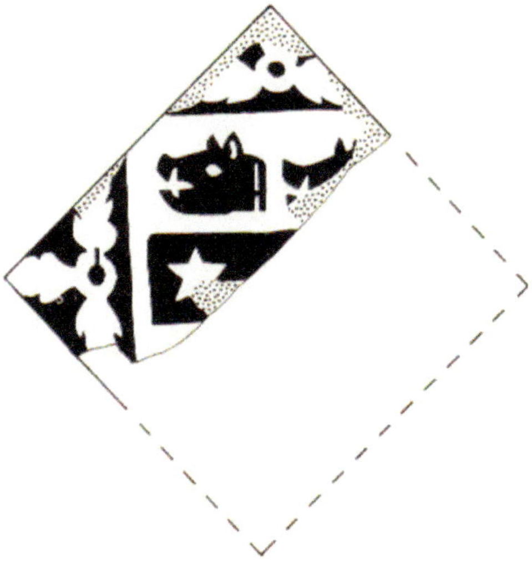

'Shield of arms set diagonally; *on a fess between three boars' heads couped as many mullets.*' Alternatively, '*a fesse between three boars' heads with three molets on the fesse*'. Surprisingly this is not recorded as present in the Vince & Wilmott paper on the lost pavement at Tewkesbury Abbey.

The royal arms are depicted on one tile; unusually not within an heraldic shield.

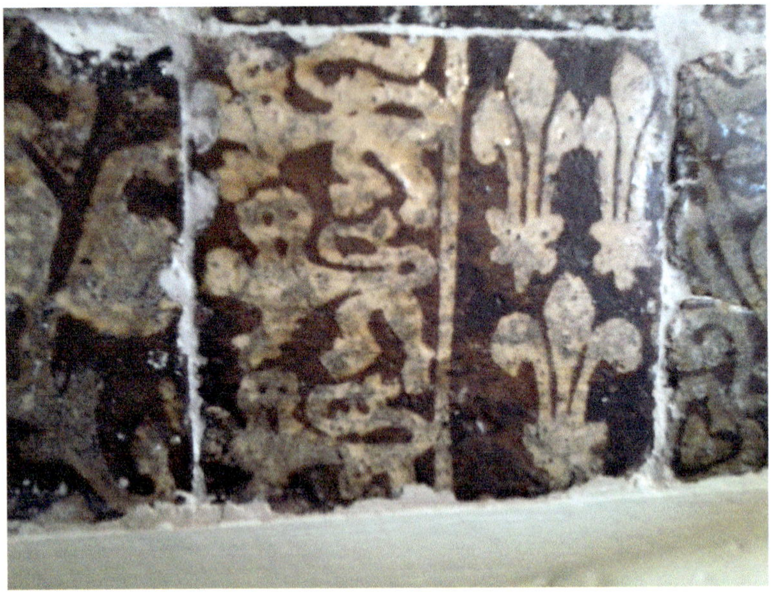

The British Museum Catalogue lists such a tile (1523) as from Bordesley Abbey in the West Midlands, and gives a possible date of thirteenth century. Dr Patrick Brown has also identified this tile within the Canynges Pavement collection at the British Museum.

Summary

The modern county of Herefordshire is clearly an artificial boundary when considering the distribution of decorated medieval floortiles. There are however some distinctive groups of designs.

The single colour relief tiles occur almost exclusively in the north of the county, centred around Wigmore and referenced in JM Lewis' book as Group 27 at Old Radnor Church. There is a strong concentration in Wigmore itself, with a few found at Dore Abbey, and a scattering at Brampton Bryan. The later two-colour tiles at Wigmore have close parallels in the Lady Chapel of Hereford Cathedral and at Worcester Cathedral.

There is then an almost duplicated set of designs at St Michael's, Croft and at Stretton Sugwas, (one with the date 1456) most of which are also recorded at Monmouth and Raglan. Excavation in Monmouth revealed a kiln site at Cadogan House in Monmouth which could conceivably have supplied these sites. Kiln sites are another topic on their own, as they can be very difficult to identify unless they are within a historical building complex. Brick kilns have been recorded and unless tile wasters are recovered, who can say what else was being fired in the kilns. I think we should be sceptical of large numbers of tiles being transported over land; it is far more likely that skilled itinerant tilers with their own set of stamps would move from place to place to fulfil commissions. Incidentally, four tile fragments from Wigmore were found fused together in pairs which would suggest the kiln site was nearby, and at Leominster, an archaeological rescue excavation across the river from the Old Priory found tile wasters of a very similar size and composition to those recovered from the Old Priory site.

In the north of the county, tiles of the same design are found at Croft, Leominster, and Brampton Bryan, and Brampton Bryan clearly has links with Wigmore.

In the west of the county, the Bristol Canynges designs occur at St Katherine's, Ledbury (probably 1486) and at Netherwood, many of which are also found at Worcester Cathedral. The unusual secular location of Deans Place at Yatton has several examples of the Canynges designs.

Dilwyn clearly has very close links to Blackfriars, Gloucester, with many of the same designs, but also what appear to be two identical damaged stamps.

Castle Frome remains the biggest puzzle with just one clearly identifiable tile of the Canynges Group. I am not aware of any similar designs to these two groups of the four-tile design with mask-like faces facing left in a circle. Compared to most of the other tiles they appear to have had little wear, and were surprisingly selected by the pious Victorians to be relaid within the aumbry. The other tiles are likely to be older and are more worn and are mostly unparalleled elsewhere in the county.

Mention should be made of other apparently original designs. Leominster Old Priory has the clearest example of an individual tiler making his stamp in the fragments which bear the impression of a key to a barrel lock. Leominster also has a chequer board design with dots on the board, not found elsewhere, and also the rare edging or border strips, apparently fired individually.

Although only a tiny percentage of the Herefordshire tiles are not in use as flooring, of the ones whose underside is visible the keying marks are usually the four or five small square stabbed impressions (like a Phillips screwdriver). They are not common and their purpose is unclear. It is suggested they aided the firing of the tiles or were utilised in the laying plan, but neither argument is convincing. There are two examples of large scoops removed from the underside, from Craswall Priory (in the Hereford Museum Collections).

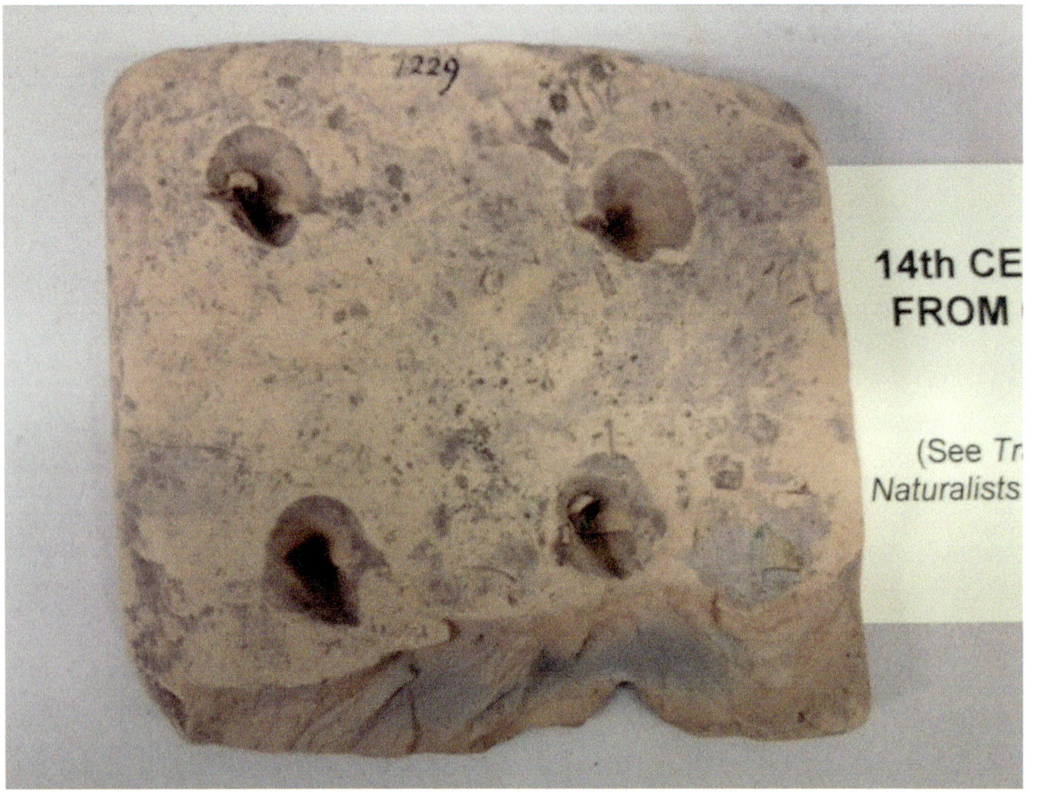

It is my sincere wish that this superficial survey of the medieval floortiles in Herefordshire will be of value to future students.

Bibliography

BIRD, T (1828) (*re Eign Gate, Hereford*) Herefordshire County Archives D.69

BORDER ARCHAEOLOGY (2002) Report on *Leominster Old Priory* in Herefordshire HER 31935 NGR SO 4980 5935

BOWEN, J. (2007) *The Medieval Floor Tiles from Leominster Priory*: unpublished BA dissertation

BRAKSPEAR, H. (1933) *(re Wigmore Abbey)* Archaeological Journal 90, 33 and pl vi

EAMES, Elizabeth A. (1980) *A Catalogue of Medieval Lead Glazed Tiles in the British Museum;* British Museum Publications, London

HILLABY, Joe and Caroline (2006*) Leominster Minster, Priory and Borough:* Logaston Press, Herefordshire

HILLABY, Joe (2003) *St Katherine's Hospital at Ledbury*: Logaston Press, Herefordshire

LEWIS, J. M. (1999) *The Medieval Tiles of Wales:* National Museum of Wales, Cardiff
(drawings by Jeremy Leech)

PEVSNER, N. (1963) *The Buildings of England: Herefordshire*: Penguin Press, London

SHOESMITH, R. and RICHARDSON, RUTH E. (eds) (1997) *A Definitive History of Dore Abbey:* Logaston Press, Herefordshire

SIDEBOTHAM, Audrey (1956) *Brampton Bryan Church and Castle*: Orphans Press Ltd, Leominster

TILEWEB.ASHMOLEAN.ORG Parker-Hore Collection

TRANSACTIONS OF THE WOOLHOPE NATURALISTS' FIELD CLUB, Herefordshire
1870 re Wigmore Abbey: 1918-20 re Brampton Bryan

WIGHT, Jane A. (1975) *Medieval Floortiles:* John Baker, London

VINCE, A.G. and WILMOTT, A. (1991) *A Lost Tile Pavement at Tewkesbury Abbey*: Antiquaries Journal LXXI, 138-173

VINCE, A.G. and SHOESMITH, R. (1998) *The Medieval and late ceramic building materials from Goodrich Castle:* AVAC Report